Evaluating Process Safety in the Chemical Industry

A USER'S GUIDE TO QUANTITATIVE RISK ANALYSIS

This is a volume in the CCPS Concepts Series.
A complete list of books available from CCPS is printed at the end of this book.

Evaluating Process Safety in the Chemical Industry

A USER'S GUIDE TO QUANTITATIVE RISK ANALYSIS

J. S. Arendt

D. K. Lorenzo

EQE International, Inc.
Knoxville, Tennessee

American Chemistry Council
1300 Wilson Boulevard
Arlington, Virginia 22209

Center for Chemical Process Safety
3 Park Avenue
New York, New York 10016

ISBN 0-8169-0746-3

Library of Congress Cataloging-in-Publication Data
(CIP data has been applied for)

This guide was prepared by EQE International, Inc. (EQE), an ABS Group Company, as an account of work sponsored by the American Chemistry Council, formerly the Chemical Manufacturers Association, and the Center for Chemical Process Safety (CCPS) of the American Institute of Chemical Engineers (AIChE). It is sincerely hoped that the information presented in this document will lead to an even more impressive record for the entire industry; however, the American Institute of Chemical Engineers, the American Chemistry Council, its consultants, CCPS Subcommittee members, their employers, their employers' officers and directors, and EQE International, Inc. and its employees disclaim making or giving any warranties or representations, express or implied, including with respect to fitness, intended purpose, use or merchantability and/or correctness or accuracy of the content of the information presented in this document. As between (1) American Institute of Chemical Engineers, its consultants, CCPS Subcommittee members, their employers, their employers' officers and directors, and EQE International, Inc. and its employees and (2) the user of this document, the user accepts any legal liability or responsibility whatsoever for the consequence of its use or misuse.

PRINTED IN THE UNITED STATES OF AMERICA
10 9 8 7 6 5 4 3 2 1

CONTENTS

LIST OF FIGURES

LIST OF TABLES

PREFACE

Quantitative risk analysis (QRA) is a powerful analysis approach used to help manage risk and improve safety in many industries. When properly performed with appropriate respect for its theoretical and practical limitations, QRA provides a rational basis for evaluating process safety and comparing improvement alternatives. However, QRA is not a panacea that can solve all problems, make decisions for a manager, or substitute for existing safety assurance and loss prevention activities. Even when QRA is preferred, qualitative results, which always form the foundation for QRA, should be used to verify and support any conclusions drawn from QRA.

The American Chemistry Council, the Center for Chemical Process Safety, and their member companies recognize the need to provide decision makers with a guide to QRA. Chemical process industry (CPI) professionals at every level (plant managers, project managers, engineers, supervisors, etc.) need criteria for determining when risk analysis will provide information that will aid their decision making. Executives need help in understanding and evaluating QRA results that are often inscrutable to nonexperts. And all need advice concerning how detailed an analysis must be if it is to provide adequate information for a specific decision. By illustrating the judicious use of QRA, this guide will help managers use their limited resources more efficiently.

This guide summarizes some of the wisdom accumulated by CPI risk analysis practitioners and safety professionals; decision makers considering the use of QRA can benefit from this collected experience. Even though this guideline does not address every issue and circumstance concerning the use of QRA, we believe that you will be able to blend your experience with the strategies provided in this guide to make more informed decisions about using QRA.

ACKNOWLEDGMENTS

The American Chemistry Council, formerly the Chemical Manufacturers Association (CMA), and the American Institute of Chemical Engineers Center for Chemical Process Safety (AIChE/CCPS) have jointly published *Evaluating Process Safety in the Chemical Industry: User's Guide to Quantitative Risk Analysis*. This is a revised and updated edition of *Evaluating Process Safety in the Chemical Industry: A Manager's Guide to Quantitative Risk Analysis*, published in 1989 by CMA.

This book was written by Donald K. Lorenzo and J. Steven Arendt of EQE International, Inc. (EQE), an ABS Group Company. The technical insights, experiences, suggestions, and peer review of the CCPS Risk Analysis Subcommittee (RASC), the American Chemistry Council Process Safety Subteam (PSS), and the original (1989) CMA Process Safety Analysis Task Group (PSATG) were essential to the development of this guide.

The CCPS RASC was chaired by Dennis C. Hendershot (Rohm & Haas Company), and committee members included Daniel A. Crowl (Michigan Technological University), Scott W. Ostrowski (Exxon Mobil Chemical), Randy Freeman (Solutia, and subsequently, EQE), William Lutz (Union Carbide), Chuck Fryman (FMC Corporation), Della Wong (NOVA Chemicals), Walter Silowka (Air Products and Chemicals, Inc.), William Tilton (DuPont), Arthur Woltman (Shell), and Thomas Gibson (CCPS).

The American Chemistry Council PSS was chaired by Peter Lodal (Eastman Chemical) and included representatives from DuPont; Shell Chemical; ExxonMobil Chemical; Huntsman; Dow Chemical; Solutia; Celanese; Rohm & Haas Company; Air Products and Chemicals, Inc.; Monsanto; Union Carbide; Montell; Eastman Chemical; and Lyondell.

We are also indebted to the reviewers of this guide at EQE: John Farquharson for his technical review, Maureen Hafford for her editorial suggestions, Leslie Adair for proofreading the guide, and Angie Nicely and Paul Olsen for preparation of the manuscript for this guide.

EXECUTIVE SUMMARY

The art of making wise decisions is the hallmark of successful management and requires both pertinent information and good judgment. Safety-related decisions, in particular, have traditionally been based on hard-earned operating experience and intuition. As greater demand for improving the safety, health, environmental, and economic aspects of facilities is placed on companies' finite resources, the decision-making process becomes more difficult and the need for better information becomes more critical.

Company management now recognizes that simply reacting to accidents and then determining where additional safety precautions are needed is no longer acceptable — the potential effects of accidents are becoming increasingly catastrophic. Moreover, today's technical and social environment dictates that decision makers take a more proactive approach to safety-related decision making and that more thorough methods and strategies be used to gain an increased understanding of the significance of risks from their companies' operations.

Risk is defined as the combination of the expected frequency and consequence of accidents that could occur as a result of an activity. ***Risk analysis*** is a formal process of increasing one's understanding of the risk associated with an activity. The process of risk analysis includes answering three questions:

- What can go wrong?
- How likely is it?
- What are the impacts?

Risk assessment is the subsequent process of taking risk analysis results and answering a fourth question:

- Are the risks tolerable?

Qualitative answers to one or more of these questions are often sufficient for making good decisions about the allocation of resources for safety improvements. But, as managers seek quantitative cost/benefit information upon which to base their decisions, they increasingly turn their attention to the use of ***quantitative risk analysis*** (QRA).

This guide provides information on the applicability of QRA to the chemical process industry (CPI). Although companies have many possible applications for risk analysis (e.g., determining the investment risk of a new product), this guide focuses on how risk analysis methods can be used for the improvement of process facilities. Moreover, while QRA can also be used to investigate economic, environmental, and health risks of process operations, this guide concentrates on QRA's use for estimating one particular type of risk—the risk of immediate injuries to workers or the public from single accidents involving acute exposure to energy releases or harmful substances.

Developing an appreciation of the benefits, limitations, relative costs, and complexities of using QRA is a necessity for CPI managers. To equip the potential user of QRA with this basic understanding, this guide discusses three important aspects of QRA:

- How to decide whether to use QRA
- How to set up a QRA to provide specific risk information
- How to interpret and use QRA results

This guide presents a framework to help you decide whether QRA can aid your decision making. Various factors influencing the decision to use QRA are described, and the types of information QRAs make available to managers are discussed. Managers are encouraged to first use qualitative techniques and risk screening methods as decision aids. Efficiency dictates that managers use QRA only in selected cases when decision-making information cannot be supplied by less elaborate means. But, appropriately scoped and applied, QRA can provide powerful insights for allocating finite process safety resources. This guide contains a flowchart of questions and information you can use to help determine when to use QRA.

If decision makers choose to use QRA, they must then define the analysis objectives so the results will satisfy the particular decision-making requirement. Because the cost of performing QRA is dependent on depth and scope of study, this guide stresses the importance of defining the right problem for analysis. An overview of QRA methods is presented to help executives understand the options available when selecting QRA techniques. To help managers have realistic expectations, important limitations of QRA techniques are also discussed.

Finally, this guide presents information on interpreting and using QRA results, outlining several methods for comparing results with experience and for presenting results to enhance credibility. Because the way people view risk is an overriding concern in the use of QRA, various factors that influence risk perception are also discussed. And the guide lists some pitfalls managers should avoid in using QRA results for decision making.

When QRA is used judiciously, its advantages can outweigh the associated problems. However, companies should resist the indiscriminate use of QRA as a means to solve all problems since this strategy could be an inefficient use of finite safety improvement resources, diverting attention from other essential safety activities. Once executives can interpret and use QRA results, they will appreciate that the quality of their decisions largely rests on their ability to understand the salient analysis assumptions. Moreover, they can use QRA to determine the impacts of important assumptions, and can use these sensitivity results to better understand the limitations of QRA studies.

Quantitative risk analysis is an important tool for the CPI. But QRA must complement (and not replace) other historically successful methods for safety assurance, loss prevention, and environmental control. A new, evolving technology, and still more of an art than a science, QRA will never make a decision for you—it can only help to increase the information base you draw on when making a decision. More conventional process safety management practices, such as good design standards, proper construction, accurate procedures, thorough training, periodic safety audits, and sound management judgment, will continue to form the foundation for a safe and productive chemical industry.

ADVICE FOR THE READER

This guide is designed to equip you with a basic understanding of the benefits, limitations, and complexities of using QRA. However, this is not a "how to" manual for QRA; nor does it concentrate on how to set up a corporate QRA program. (This information can be found in the CCPS *Guidelines for Chemical Process Quantitative Risk Analysis, Second Edition.*[24]) Instead, this guide describes the role managers and sponsors should play in ensuring the success of QRA projects. To convey this information, we use the following steps:

- Establish a basic vocabulary (Glossary). Every discipline has its own jargon, and QRA is no different
- Define a method for determining whether QRA can (or is needed to) help your decision making
- Describe what to reasonably expect from QRA
- Provide a basis for understanding QRA results, beyond the obvious statistical meanings

This guide may be read by an audience ranging from middle managers to senior executives who have different levels of knowledge about QRA. This guide may also be read by engineers and other technical staff who will be contributing to QRAs or using the results. We have, therefore, designed the chapters to allow for differences in expertise and need.

Chapter 1 defines QRA, discusses its essential elements, and dispels some misconceptions. Chapter 2 outlines considerations for deciding when to apply QRA. It presents some reasons for performing QRA and describes the types of information available from such studies. This chapter also describes practical situations in which QRA may be used successfully, as well as conditions that make QRA a less desirable choice.

Once the decision has been made to use QRA, the next step is to execute it effectively. Chapter 3 describes the process of setting up an individual QRA. This chapter discusses the importance of defining the right problem for analysis and selecting the right analysis techniques; it also provides an overview (not a how to) of the various classes of QRA techniques. Chapter 4 discusses ways to interpret and use QRA results. Conclusions about the future of QRA in the CPI are offered in Chapter 5.

ACRONYMS

AIChE	American Institute of Chemical Engineers
API	American Petroleum Institute
CCPS	Center for Chemical Process Safety
CMA	Chemical Manufacturers Association
CPI	Chemical process industry
CPQRA	Chemical process quantitative risk analysis
ERPG	Emergency response planning guideline
FAR	Fatal accident rate
FMEA	Failure modes and effects analysis
HAZOP	Hazard and operability analysis
LOPA	Layer of protection analysis
P&ID	Piping and instrumentation diagram
PRA	Probabilistic risk analysis
QRA	Quantitative risk analysis
RASC	Risk Assessment Subcommittee
ROD	Average rate of death
STEL	Short-term exposure limit

GLOSSARY

Acceptable risk	The average rate of loss that is considered tolerable for a given activity
Accident (sequence)	A specific combination of events or circumstances that leads to an undesirable consequence
Acute hazard	The potential for injury or damage to occur as a result of an instantaneous or short duration exposure to the effects of an accident
Aggregate risk	Societal risk for onsite workers in occupied buildings (API 752)
Average aggregate risk	Average societal risk for onsite workers in occupied buildings (API 752)
Average individual risk	Average individual risks must be defined in context of the exposed population: A. ***Average individual risk (exposed population)*** is the individual risk averaged over the population that is exposed to risk from the facility B. ***Average individual risk (total population)*** is the individual risk averaged over a predetermined population, without regard to whether or not all people in that population are actually exposed to the risk C. ***Average individual risk (exposed hours/worked hours)*** for an activity may be calculated for the duration of the activity or may be averaged over the working day
Average rate of death (ROD)	The average number of fatalities that might be expected per unit time from all possible incidents
Chronic hazard	The potential for injury or damage to occur as a result of prolonged or repeated exposure to an undesirable condition
Consequence	The direct, undesirable result of an accident, usually measured in health/safety effects, environmental damage, loss of property, or business costs

CPQRA — The acronym for chemical process quantitative risk analysis. It is the process of hazard identification followed by numerical evaluation of incident consequences and frequencies, and their combination into an overall measure of risk when applied to the chemical process industry. It is particularly applied to episodic events. It differs from, but is related to, a probabilistic risk analysis (PRA), a quantitative tool used in the nuclear industry

Dispersion model — A mathematical model describing how material is transported and dispersed from a release

Emergency response planning guideline (ERPG) — A system of guidelines for air concentrations of toxic materials prepared by the American Industrial Hygiene Association. An ERPG is the maximum airborne concentration below which it is believed that nearly all individuals could be exposed for up to 1 hour with the following results:

- ERPG-1: Nothing other than mild transient adverse health effects or perception of a clearly defined objectionable odor
- ERPG-2: No irreversible or other serious health effects or symptoms that could impair an individual's ability to take protective action
- ERPG-3: No life-threatening health effects

Episodic event — A release of limited duration, usually associated with an accident

Event tree (analysis) — A logic model that graphically portrays the range of outcomes from the combinations of events and circumstances in an accident sequence. For example, a flammable vapor release may result in a fire, an explosion, or in no consequence depending on meteorological conditions, the degree of confinement, the presence of ignition sources, etc. These trees are often shown with the probability of each outcome at each branch of the pathway

Expected value — The statistical average of a variable described by a probability distribution

Failure modes and effects analysis (FMEA) — A hazard identification technique in which all known failure modes of components or features of a system are considered in turn, and undesired outcomes are noted

Fatal accident rate (FAR) — The average number of fatalities expected in a particular worker population of interest over a period of 10^8 worker-hours (roughly 1,000 employee working lifetimes)

Fault tree (analysis) — A logic model that graphically portrays the combinations of failures that can lead to a particular main failure (TOP event) or accident of interest. Given appropriate data, fault tree models can be quantitatively solved for an array of system performance characteristics (mean time between failures, probability of failure on demand, etc.)

F-N curve	A graphical illustration of the cumulative frequency (F) of accidents resulting in a consequence of greater than or equal to N impacts. A way of illustrating societal risk
Frequency	Number of occurrences of an event per unit of time
Hazard	A chemical or physical condition that has the potential for causing damage to people, property, or the environment
Hazard and operability (HAZOP) analysis	A technique to identify process hazards and potential operating problems using a series of guide words to study process deviations
Importance	The probability a specific component (or collection of components) is contributing to a system failure, given that the system is failed
Incident	An unplanned release of hazardous chemicals or energy
Individual risk	The risk to a person in the vicinity of a hazard. This includes the nature of the injury to the individual, the likelihood of the injury occurring, and the time period over which the injury might occur
Layer of protection analysis (LOPA)	A simplified form of event tree analysis using selected accident scenarios and order-of-magnitude estimates to determine whether additional protection is needed
Likelihood	A measure of the expected probability or frequency of occurrence of an event
Probability	The expression for the likelihood of occurrence of an event or an event sequence during an interval of time or the likelihood of the success or failure of an event on test or on demand. By definition, probability must be expressed as a number ranging from 0 to 1
Process safety management	A program or activity involving the application of management principles and analytical techniques to ensure the safety of chemical process facilities
Quantitative risk analysis (QRA)	The systematic development of numerical estimates of the expected frequency and/or consequence of potential accidents associated with a facility or operation
Rare event	An event or accident whose expected frequency is very small. The event is not expected to occur during the normal life of a facility or operation
Risk	A measure of human injury, environmental damage, or economic loss in terms of both the incident likelihood and the magnitude of the loss or injury

Risk analysis The development of a quantitative estimate of risk based on engineering evaluation and mathematical techniques for combining estimates of incident consequences and frequencies

Risk assessment The process by which the results of a risk analysis (i.e., risk estimates) are used to make decisions, either through relative ranking of risk-reduction strategies or through comparison with risk targets

Risk contour Lines that connect points of equal risk around the facility ("iso-risk" lines)

Risk estimation Combining the estimated consequences and likelihood of all incident outcomes from all selected incidents to provide a measure of risk

Risk management The systematic application of management policies, procedures, and practices to the tasks of analyzing, assessing, and controlling risk in order to protect employees, the general public, and the environment, as well as company assets, while avoiding business interruptions

Risk measures Ways of combining information on likelihood with the magnitude of loss or injury (e.g., risk indices, individual risk measures, and societal risk measures)

Risk targets Objective-based risk criteria established as goals or guidelines for performance

Societal risk A measure of risk to a group of people. It is most often expressed in terms of the frequency distribution of multiple casualty events

Uncertainty A measure, often quantitative, of the degree of doubt or lack of certainty associated with an estimate of the true value of a parameter

1

INTRODUCTION

What the decision maker
wants is access to hope.
G.L.S. SHACKLE

1.1. BACKGROUND

Successfully managing industrial facilities requires pertinent information and good judgment. When you must make a decision affecting the level of safety of your organization's various enterprises, you need information about the risks posed by the activities of interest. Once in possession of these risk insights, you can be more effective in making risk management decisions. If information concerning the risk impact of possible choices is not available, then you are less likely to make an optimal decision.

Historically, managers in the chemical process industry (CPI) have relied upon industry experience when judging the risks associated with their facilities and activities.[1,2] The CPI has been successful in maintaining an excellent safety record compared to industry overall. But as new process technologies are developed and deployed, less of the historical experience base remains pertinent to safety assurance. Other potentially hazardous industries—such as nuclear power, aerospace, and defense—have lacked the prior experience necessary to assess the safety aspects of the advanced technology of new designs.[3,4] The absence of relevant historical data in these industries led to the development of techniques for predicting risks, including many of these now used to perform quantitative risk analysis (QRA).[5] The CPI has adapted many of these techniques and has developed new methods to deal with the diverse hazards of chemical process facilities.

QRA is fundamentally different from many other chemical engineering activities (e.g., chemistry, heat transfer, reaction kinetics) whose basic property data are theoretically deterministic. For example, the physical properties of a substance for a specific application can often be established experimentally. But some of the basic "property data" used to calculate risk estimates are probabilistic variables with no fixed values. Some of the key elements of risk, such as the statistically expected frequency of an accident and the statistically expected consequences of exposure to a toxic gas, must be determined using these probabilistic variables. QRA is an approach for estimating the risk of chemical operations using the probabilistic information. And it is a fundamentally different approach from those used in many other engineering activities because interpreting the results of a QRA requires an increased sensitivity to uncertainties that arise primarily from the probabilistic character of the data.

Estimating the frequencies and consequences of rare accidents is a synthesis process that provides a basis for understanding risk. (Throughout the published literature, the terms risk assessment and risk analysis are used interchangeably in reference to this process.) Using this synthesis process, you can develop risk estimates for hypothetical accidents based upon your experience with the individual basic events that combine to cause the accident. (Basic events typically include process component failures, human errors, and changes in the process environment, and more information is usually known about these basic events than is known about accidents.) System logic models are used to couple the basic events together, thus defining the ways the accident can occur.

With the advent of this new safety analysis technology, and the need for providing better input to risk management and safety improvement decisions, many CPI safety professionals are calling for increased use of QRA. And, given the contemporary technical and social environment, *it is imperative that management personnel understand the strengths and weaknesses of QRA technology.*

1.2. THE PROCESS OF RISK ANALYSIS

Risk analysis is the process of gathering data and synthesizing information to develop an understanding of the risk of a particular enterprise. Risk analysis usually involves several of the five risk management activities shown in Figure 1. CPI companies have many possible applications

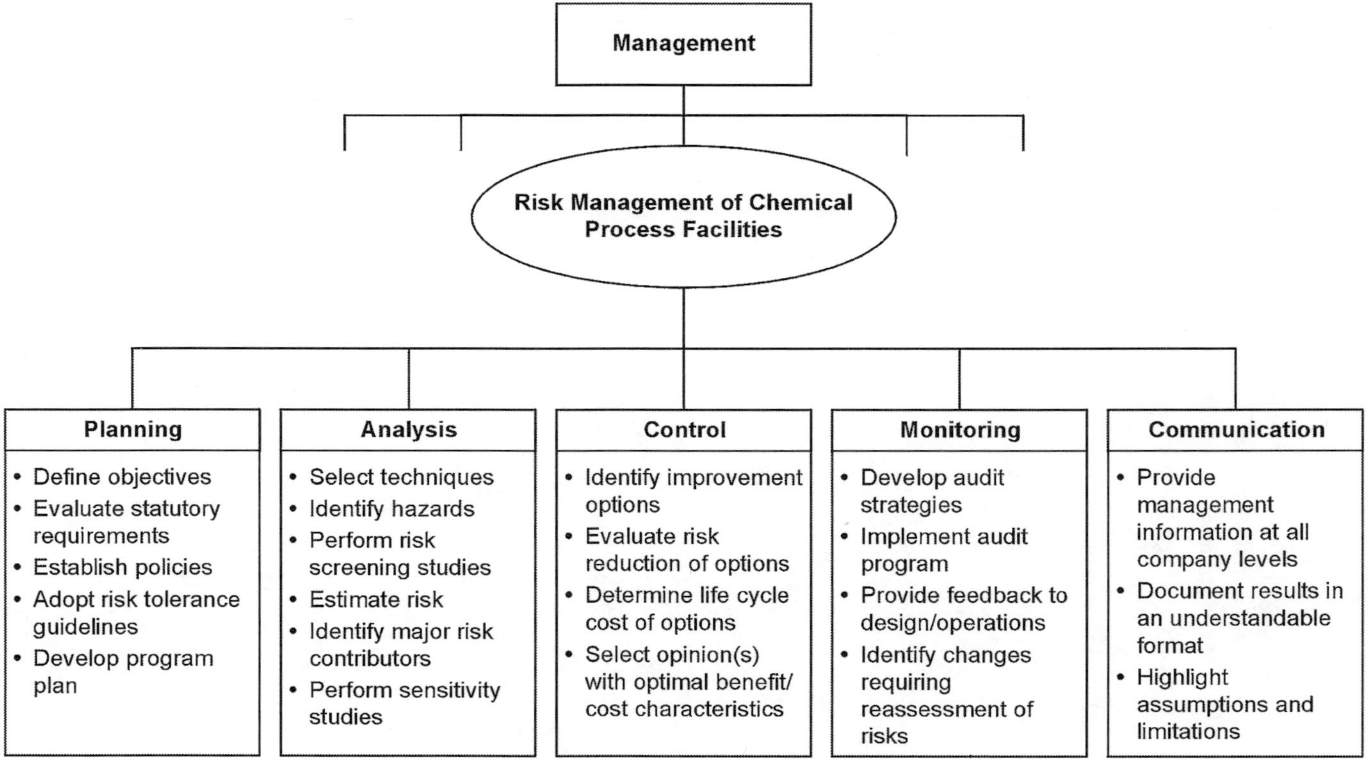

Management

Risk Management of Chemical Process Facilities

Planning
- Define objectives
- Evaluate statutory requirements
- Establish policies
- Adopt risk tolerance guidelines
- Develop program plan

Analysis
- Select techniques
- Identify hazards
- Perform risk screening studies
- Estimate risk
- Identify major risk contributors
- Perform sensitivity studies

Control
- Identify improvement options
- Evaluate risk reduction of options
- Determine life cycle cost of options
- Select opinion(s) with optimal benefit/ cost characteristics

Monitoring
- Develop audit strategies
- Implement audit program
- Provide feedback to design/operations
- Identify changes requiring reassessment of risks

Communication
- Provide management information at all company levels
- Document results in an understandable format
- Highlight assumptions and limitations

FIGURE 1. Elements of risk management.

for risk analysis.[6–11] For example, before proceeding with full-scale development of a new product, management may wish to determine whether the marketing of that product will succeed. In another instance, company executives may want to know how to allocate resources to minimize the chance of a catastrophic accident at a chemical process facility. This guide is concerned with the latter situation—assessing the risk of episodic events. With the understanding available from such risk analyses, you will be better equipped to evaluate and select risk management options.

The effort needed to develop this understanding will vary depending upon the foundation of information you have for understanding the significance of potential accidents (Figure 2). If you have a great deal of pertinent, closely related experience with the activity you wish to know the risk of, then very little formal analysis may be needed. However, even minor changes can radically increase the risk of an accident. History is replete with examples of design "improvements" or minor extrapolations that pushed a proven design beyond safe limits. If, on the other hand, there is no relevant experience base for extrapolation, you will have to rely on analytical techniques or on your own intuition for answering risk analysis questions. However, no risk analysis technique can provide meaningful results if you do not have fundamental knowledge of process hazards.

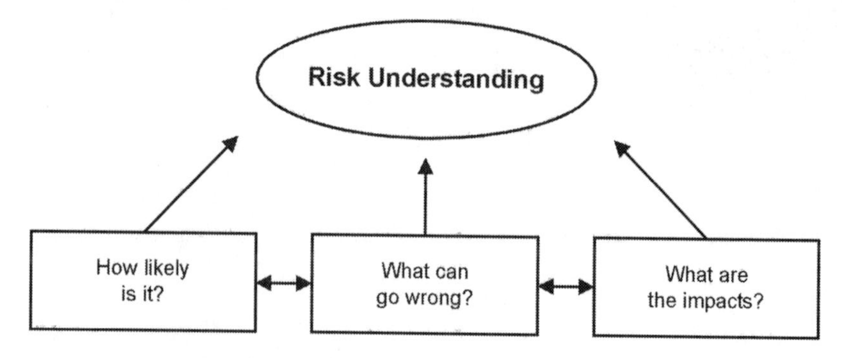

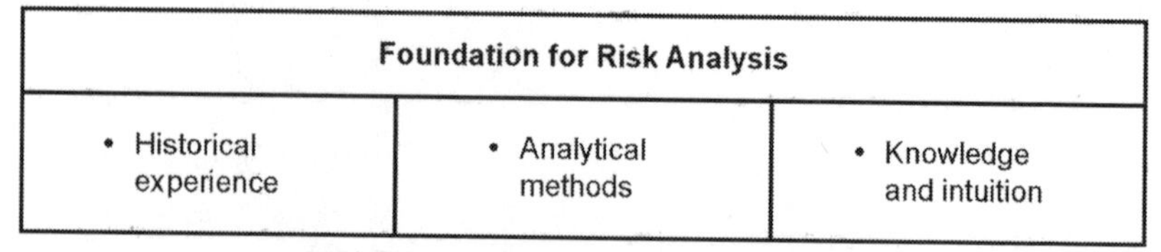

FIGURE 2. Elements of risk analysis.

If your risk understanding is inadequate, you can use the ***process of risk analysis*** (Figure 3) to acquire the understanding you need. The extent of risk analysis and the degree of risk understanding that are needed may vary. Sometimes, simply knowing what could go wrong (hazard identification) may be sufficient for your decision, and no elaborate quantification of likelihoods or effects would be needed. Occasionally, you may have sufficient understanding about what can go wrong and what the effects of an accident could be; however, you may still need information on how likely the accident is. In other cases the quantification of potential impacts alone will be adequate, and analysis of the likelihoods is unnecessary. In practice, few decisions require explicit quantification of both frequency and consequence.

1.3. DEFINITION OF QRA

QRA is the art and science of developing and understanding numerical estimates of the risk (i.e., combinations of the expected frequency and consequences of potential accidents) associated with a facility or operation. It uses a set of highly sophisticated, but approximate, tools for acquiring risk understanding. QRA methods can be used throughout all phases of the life of a process (laboratory development, detailed design, operation, demolition, etc.). However, QRA is most effective when used to analyze a process whose design characteristics have been specified (i.e., the piping and instrument diagrams [P&IDs] are available, chemical reactions and other unit operations are known, and the process control strategy is defined) and for which there exists some relevant operating experience from similar systems.

QRA can be used to investigate many types of risks associated with chemical process facilities, such as the risk of economic losses or the risk of environmental impact. But, in health and safety applications, the use of QRA can be classified into two categories:

1. Estimating the long-term risk to workers or the public from chronic exposure to potentially harmful substances or activities
2. Estimating the risk to workers or the public from episodic events involving a one-time exposure to potentially harmful substances or activities

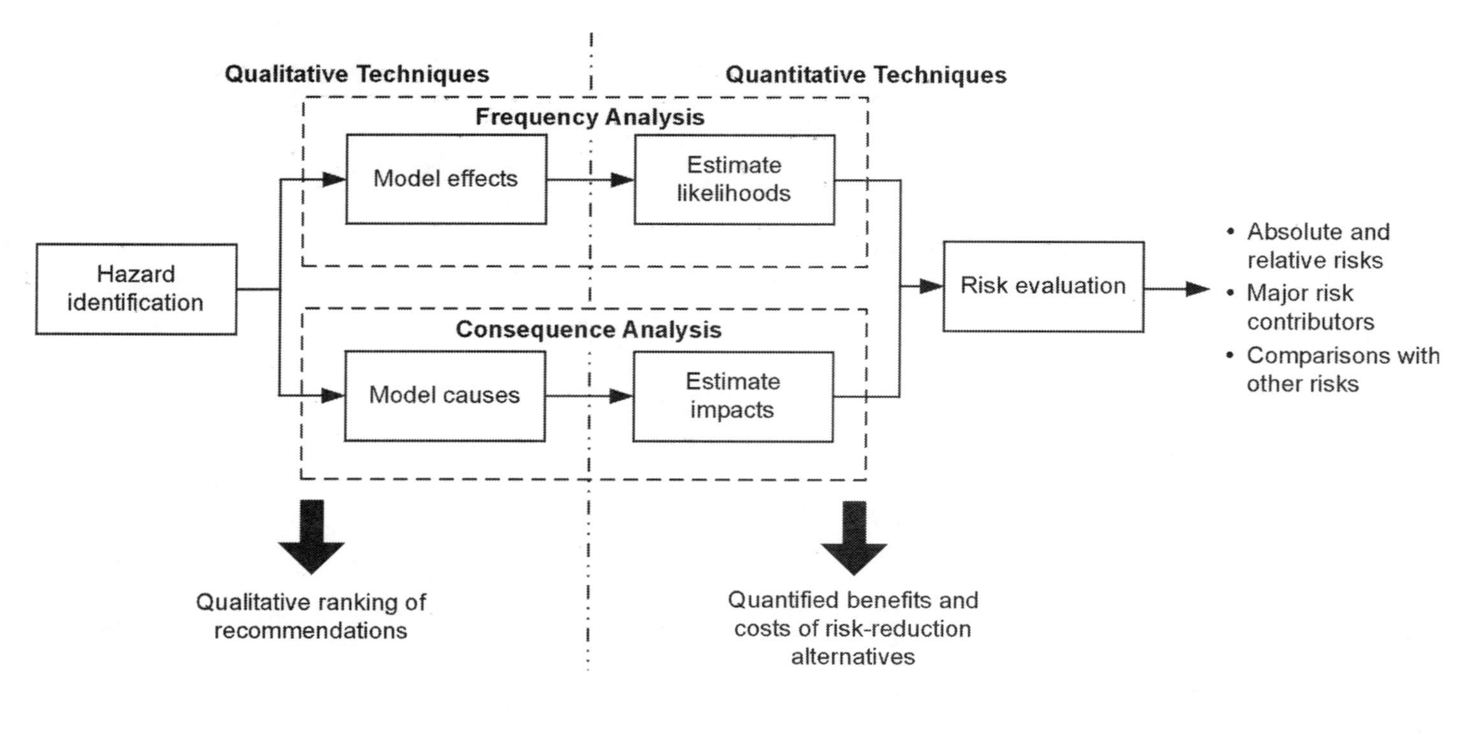

FIGURE 3. The process of risk analysis.

In this guide, we will focus on the use of QRA in the safety assessment of acute hazards and episodic events only.

1.4. MISCONCEPTIONS ABOUT QRA

Table 1 shows prevalent examples of misconceptions about QRA. Many are actually generalizations that are too broadly applied. Two of the most common misconceptions concern (1) the lack of adequate equipment failure data and (2) the cost of performing QRA.

The core function of QRA is to provide information for decision making. QRA results in and of themselves cannot prove anything. However, decision makers can compare QRA risk estimates to their own risk tolerance criteria to decide whether a plant or operation is safe enough. The same QRA results can support both the plant manager's contention that the plant is safe, as well as the community activist's claim that the plant is unsafe. The difference lies in the individual's risk tolerance, not the QRA.

Unfortunately, even if everyone agrees on a tolerable risk value, there are many other subjective factors that influence our understanding (and tolerance) of risk. If 1 fatality per year were tolerable from causes such as falls, electrocutions, or asphyxiations, would 100 fatalities be

TABLE 1. Misconceptions About QRA Technology and Risk

- A QRA can prove that the plant is safe or unsafe.
- QRA is a totally objective way to understand risk.
- If we could measure risk accurately, our decisions would be easy.
- If we do a QRA, we can reduce our risk to zero.
- We analyzed all possible accidents.
- QRA is expensive.
- QRA is cheap.
- We can usually predict risk to an accuracy of a factor of 2 or better.
- We don't have enough data to do QRA.
- We have enough data so we don't need to do QRA.
- QRA is pure science.

equally tolerable from catastrophic explosions predicted to occur, on average, once every 100 years? In both cases, QRA results would predict an average risk of one fatality per year. Are worker injuries more tolerable than public injuries? Are injuries to adults more tolerable than injuries to children? Typical QRA results simply report risk as injuries per year. Yet, as discussed in Section 4.2, there are many other subjective factors that influence a decision maker beyond the "objective" numeric results of a QRA.

No matter how accurate the QRA results are, the conscious decision to accept risk (actually, the decision is whether to spend more money to further reduce the risk) is always difficult when *near the risk tolerance threshold*. If the risk is clearly above tolerable thresholds (e.g., the risk of fire in a flammable storage area if uncontrolled welding operations are performed), then the decision to spend money to reduce that risk (e.g., to install a fire suppression system, to train a fire brigade, or to implement a hot work permit system) is relatively easy. Similarly, if the risk is clearly small (e.g., the risk of a meteorite puncturing a tank), then the decision to spend no money on meteorite shields is equally easy. However, should a high-high pressure alarm be installed in addition to the existing high pressure alarm and relief valve? The QRA results show that the change would reduce risk, but the manager must decide whether the benefit is worth the cost.

QRA results can guide decision makers in their quest for continuous improvement in risk reduction, but zero risk is an unattainable goal. Any activity involves some risk. Even if it were hypothetically possible to eliminate the risk of every accident scenario in a QRA, some risk would still remain because no QRA examines every possible accident scenario. At best a QRA identifies the dominant contributors to risk from the system as it existed at the time of the analysis. Once those are eliminated, other minor risk contributors (including many that were left out of the original QRA because they were "negligible" contributors, as well as new risks introduced by changes to eliminate the original risks) remain as the new dominant risk contributors.

The availability of resources to perform the analysis is the primary constraint on the completeness of QRAs. Managers must balance the value of QRA results in their decision making against the cost of obtaining these results. It has been shown repeatedly that, when properly scoped and executed, QRA is very cost-effective. In the past, QRA has been used with little regard for minimizing analysis cost versus benefit (e.g., in the nuclear power industry). But QRA can be cost-effective when appropriately preceded by qualitative evaluations and risk screening methods[12, 13] that reduce the size and complexity of the QRA study.

The accuracy of QRA results is also dependent on the analysis resources. Obviously, more complete QRA models can produce more accurate results. But even the best model is worthless if the input data are speculative or erroneous. Fortunately, the scarcity of process-specific data for some applications may not be an insurmountable problem. There exist a few industrywide databases, such as those in Table 2, that

TABLE 2. Example Data Sources

The *Guidelines for Process Equipment Reliability Data with Data Tables*[14] covers a variety of components used in the chemical process industry, including electrical equipment, analyzers, instrumentation and controls, detectors, heat exchangers, piping systems, rotating equipment (pump, compressor, and fan), valves, and fire protection systems.	The *IEEE Guide to the Collection and Presentation of Electrical, Electronic, Sensing Component, and Mechanical Equipment Reliability Data for Nuclear Power Generating Stations* (IEEE Std. 500-1984)[15] compiles data from over a dozen other references and includes information for most types of components.
The *OREDA Offshore Reliability Data Handbook*[16] covers a variety of components used in offshore oil drilling and platforms, including gas/fire detection systems, process alarm systems, firefighting systems, pressure relieving systems, general alarm and communication systems, evacuation systems, process systems (vessels, valves, pumps, heat exchangers, and compressors), electrical and utility systems, and drilling equipment.	*Nonelectronic Parts Reliability Data 1991*[17] (NPRD-91) and *Failure Mode/Mechanism Distributions 1991*[18] (FMD-91) provide failure rate data for a wide variety of component (part) types, including mechanical, electromechanical, and discrete electronic parts and assemblies. They provide summary failure rates for numerous part categories by quality level and environment.
The *Systems Reliability Service Data Bank*[19] was set up in the 1960s to provide engineers with reliability information for complex systems. This comprehensive data bank contains information on most component types in a variety of uses in different industries.	*Nuclear Plant Reliability Data System: Annual Reports of Cumulative System and Component Reliability for Period from July 1, 1974, through December 31, 1982*,[20] serves as a source of engineering and failure statistics for the nuclear industry. It contains data for most components used in nuclear power plants.

QRA practitioners can use to satisfy some QRA objectives. Also, the American Institute of Chemical Engineers (AIChE) has sponsored a project to expand and improve the quality of component failure data for chemical industry use.[14] And many process facilities have considerable equipment operating experience in maintenance files, operating logs, and the minds of operators and maintenance personnel. These data can be collected and combined with industrywide data to help achieve reasonable QRA objectives. However, care must be exercised to select data most representative of your specific system from the wide range available from various sources. Even data from your own plant may have to be modified (sometimes by a factor of 10 or more) to reflect your plant's current operating environment and maintenance practices.

The data needed for consequence calculations are just as vital as for frequency calculations. Data are needed for source term calculations (e.g., hole size, elevation), dispersion calculations (e.g., wind direction, wind speed, vapor density, surface roughness), fire and explosion modeling (e.g., time to ignition, degree of confinement), and effects modeling (e.g., dose-response curves). Here again, there are many potential sources of data (e.g., airport weather data if you do not have onsite data) that can be applied, with appropriate care and adjustments, to your specific situation. Even when process-specific or site-specific data are sparse, QRA analysts can often use good engineering judgment to successfully compare the relative risks between design alternatives for specific process safety decisions. Thus, apparent lack of data alone should not be a "showstopper" for potential users of QRA.

Finally, decision makers must appreciate that QRA is a practical engineering art, not a pure science. Just as mechanical engineers include "fouling factors" in calculating the required surface area of heat exchangers or structural engineers include "safety factors" in calculating the required dimensions of beams, QRA analysts use engineering judgment in developing their models and in performing their calculations. Thus, decision makers must appreciate that QRA results are as dependent on the skill of the analyst as they are on the data that went into the analysis.

2

DECIDING WHETHER TO USE QRA

He who chooses the beginning of a road
chooses the place it leads to.
HARRY E. FOSDICK

Why perform QRA? There may be many reasons, but the following are two of the more prevalent ones. First, you choose to use QRA because you believe you will gain a better understanding of risk that will aid decision making. Qualitative approaches may have been tried and found inadequate for the particular application. And sometimes QRA may be the only way of obtaining a sufficient understanding of risk.

A second possibility is that, in some cases, QRA may be required by law, so you choose to do one (or several) to see what QRA is like. Some countries have for a number of years required QRA as a prerequisite to industrial expansion. Siting decisions, process selection, number of safety systems, and so forth, often are prescribed by government authorities statutorily committed to the use of QRA. In the United States, several government agencies use risk analysis on a broad scale.[6, 12, 21] The Environmental Protection Agency's risk management program regulation 40 CFR 68 requires consequence analyses of worst-case and alternative accidental release scenarios. So, to be able to discuss when QRA may be beneficial, it is necessary to investigate the process for deciding when (or when not) to use it.

2.1. SOME REASONS FOR CONSIDERING QRA

The decision to use QRA to satisfy a particular purpose may be the result of many compounding circumstances. There is no single way that the choice is made, but generally the decision-making process follows the sequence of events shown in Figure 4.

An ***underlying motivation*** triggers one or more concerns about a company's facility or activity. Sometimes the underlying motivation is simply a perception that a problem exists. A single, memorable catastrophe can also galvanize concern. The motivation of an increasing number of companies to use QRA is a proactive desire to improve safety.

The ***concerns*** generated from an underlying motivation are often related and inevitably involve safety and economic issues. The concerns coupled with internal and external ***activators*** may energize management to increased action, and these activators establish a ***need*** for greater risk understanding. Most often the need is for insight to use in making a decision. Increasingly, an additional need is to satisfy a statutory or legal obligation. And sometimes the need for considering a QRA may be to satisfy a special purpose requirement—such as information to provide to a local emergency planning committee to support their development of contingency plans for evacuations in the event of a chemical release emergency.

Whatever the need, once established it defines the ***information requirement*** that can then be the focal point from which the question of using QRA can be considered: Can QRA satisfy the information requirement in an efficient, appropriate manner? If so, all the factors that led to the decision to use QRA now become factors that help shape the objectives and scope for the particular QRA study.

2.2. TYPES OF INFORMATION AVAILABLE FROM RISK STUDIES

The reasons (i.e., the motivations, concerns, activators, and needs) for considering the use of QRA define the requirements for information. The next question is, can QRA supply the appropriate information to satisfy the need? By definition, QRA studies generate numerical estimates of the expected frequency and/or consequence(s) of undesired events. The results of the QRA can be formulated and used on two bases: (1) an absolute basis and (2) a relative basis.

Absolute risk results are specific numerical estimates of the frequencies and/or consequences of process facility accidents synthesized from

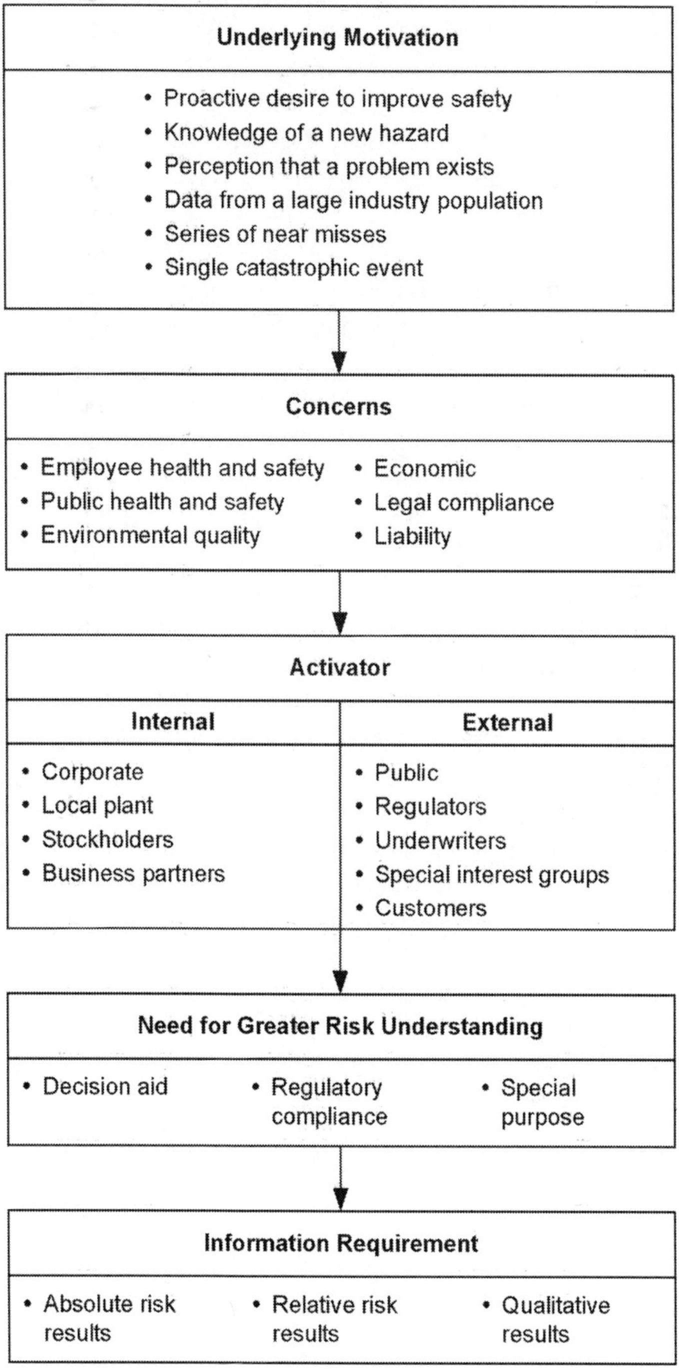

Underlying Motivation

- Proactive desire to improve safety
- Knowledge of a new hazard
- Perception that a problem exists
- Data from a large industry population
- Series of near misses
- Single catastrophic event

↓

Concerns

- Employee health and safety
- Public health and safety
- Environmental quality
- Economic
- Legal compliance
- Liability

↓

Activator

Internal	External
• Corporate • Local plant • Stockholders • Business partners	• Public • Regulators • Underwriters • Special interest groups • Customers

↓

Need for Greater Risk Understanding

- Decision aid
- Regulatory compliance
- Special purpose

↓

Information Requirement

- Absolute risk results
- Relative risk results
- Qualitative results

FIGURE 4. The evolution of a decision to use QRA.

accident models and basic input data. Theoretically, absolute risk estimates can be used to determine whether the level of safety at a facility meets risk tolerance criteria. If it does not, then changes to the facility can be made to lower the risk until it meets the risk tolerance criteria. In this sense absolute risk estimates are designed to answer the question, is the plant safe enough?

Relative risk results show only the difference between the levels of safety of one or more cases of interest and a reference, or baseline, case. Relative risk estimates can be used (as can absolute estimates) to determine the most efficient way to improve safety at a facility. But, the use of relative risk estimates alone does little to help ensure that the most efficient way is safe enough unless one of the cases meets qualitative safety criteria (e.g., compliance with relevant codes, standards, and/or regulations; consistency with current industry practice).

There are a variety of ways to express absolute QRA results. Absolute frequency results are estimates of the statistical likelihood of an accident occurring. Table 3 contains examples of typical statements of absolute frequency estimates. These estimates for complex system failures are usually synthesized using basic equipment failure and operator error data. *Depending upon the availability, specificity, and quality of failure data, the estimates may have considerable statistical uncertainty* (e.g., factors of 10 or more because of uncertainties in the input data alone). When reporting single-point estimates or best estimates of the expected frequency of rare events (i.e., events not expected to occur within the operating life of a plant), analysts sometimes provide a measure of the sensitivity of the results arising from data uncertainties.

TABLE 3. Examples of Absolute Frequency Estimates

- The expected frequency of having an explosion in the plant is 5×10^{-4} per year.
- We expect that four large toxic releases will occur during the lifetime of this facility.
- The probability of a large release of chlorine sometime during a 1-year period is 2×10^{-3}.
- The probability of safety system failure is 4×10^{-4} per batch.
- We expect to see, on the average, one small fire every month in this process building.
- The mean time between runaway reactions in this reactor is 1000 years.

Sometimes the expected consequences of an accident alone may provide you with sufficient information for decision-making purposes. Conventionally, the form of these estimates will be dictated by the purpose (concern) of the study (safety, economics, etc.). Absolute consequence estimates are best estimates of the impacts of an accident and, like frequency estimates, may have considerable uncertainty. Table 4 contains examples of typical consequence estimates obtained from QRA. These examples point to the difficulty in comparing various safety and economic results on a common basis—there is no common denominator.

If both frequency and consequence values are calculated and reported on an absolute basis, then they may be reported graphically in combination with one another (Chapter 3), or simply as the product of frequency and consequence. Table 5 contains some examples of typical risk estimates (frequency and consequence products). Based on absolute risk estimates, you can decide whether the risk of a specific activity exceeds your threshold of risk tolerance (risk goal). If so, analysts can estimate the reduction in risk, given that certain improvements are made, assumptions changed, or operating circumstances eliminated.

TABLE 4. Examples of Absolute Consequence Estimates

- This accident has the potential to seriously injure 50 people because of blast overpressure and thermal radiation effects.
- If this event occurs, we expect the process to sustain $2 million in equipment loss and 3 months of downtime.
- The maximum downwind center line concentration of hydrogen fluoride beyond the plant boundary is estimated to be 500 ppm, given that the release occurs.
- If the reactor detonates, we estimate that 20 employee fatalities will occur, and 50 members of the public will be hurt.
- The toxic plume is expected to extend 4000 meters downwind at concentrations above the short-term exposure limit (STEL).
- The results indicate that 2000 people will be exposed to a concentration of ammonia greater than the emergency response planning guideline concentration (e.g., ERPG-2).
- If the pipe breaks, we expect a 100 kg-per-second release of butane into the diked area.
- The maximum distance that a 1-psi overpressure will be felt is estimated to be 500 meters.

TABLE 5. Examples of Absolute Risk Estimates

- The risk to employees from this process is 5×10^{-4} expected fatalities per year.
- The annual economic risk of operating this unit is estimated to be $1 million because of fire and explosion accidents.
- This analysis shows that less than 1 injury per year is expected, but the frequency of injuring 100 or more people is once every 300 years, and the frequency of injuring 1000 or more people is once every 5000 years.
- We calculate the frequency of accident A as once every 5 years and accident B as once every 1000 years. The total loss if A occurs will be $1 million. The total loss if B occurs will be $200 million. The risk of A and B is the same—$200,000 per year.

Then, the absolute reduction in frequency, consequence, or risk can be calculated and compared to the cost of implementing the improvement, allowing you to determine whether the change represents the best use of resources to improve safety.

The advantage of absolute risk estimates is their ability to tell the decision maker when certain safety improvements are no longer an efficient use of resources. Conceptually, they can be used as demarcations—if the risk numbers are above the limit, you expend resources until you get the numbers below the limit. The disadvantages of using absolute estimates in this context are (1) you can never be certain about the accuracy of the results, (2) there are no standard criteria for risk tolerance that everyone agrees on for all circumstances, and (3) the numerical estimates are difficult for nonexperts to interpret. Furthermore, this approach does not consider the resources required to achieve the risk goal. Vast sums may be spent to reduce one type of risk to achieve a target, but those same resources might have been better spent reducing other risks. Senior management must take a mature and cautious approach to using absolute risk estimates in the decision-making process. Quantified risks are not necessarily the dominant risks, and obsession with a few specific estimates may distort your resource allocation so that you fail to achieve your true objective, such as saving the most lives per dollar invested in risk reduction.

The advantage of using relative risk results is that you can decide on the best way to improve safety at a facility without having to defend the absolute accuracy of the results. Relative results are also much less likely to be misinterpreted by people unfamiliar with QRA. The disadvantage of using relative results is that they, by definition, cannot give direct advice on when to stop making improvements. Table 6 contains some examples of relative estimates obtained from QRA.

There are several ways to produce relative risk estimates. One way is to calculate the risk estimates of a datum or baseline case and use them to normalize the absolute estimates for other analysis cases. Consider the following example where managers compare the risks of three process designs in order to pick the best system for manufacturing a particular chemical product. The risk estimates (the expected number of fatalities per year associated with the operation of each system) for each system are listed in the middle column of Table 7. Using System A as the baseline case, the risk of System B and System C can be compared with System A in the following manner. Define a risk index as the quotient of the risk of any option to the risk of System A. Thus, as listed in the right-hand column of Table 7, the risk index for System A is 1, the risk index for System B is 0.25 (i.e., $2 \times 10^{-5}/8 \times 10^{-5}$), and the risk index for System C is 5 (i.e., $4 \times 10^{-4}/8 \times 10^{-5}$). In other words System B presents one-fourth the risk of System A, and System C presents *five* times more risk than System A. The managers in this example could use this information,

TABLE 6. Examples of Relative Risk Estimates

- The risk from process A is about 15 times greater than the risk from process B.
- A safety integrity level 2 interlock is required to reduce the risks of the alternatives to comparable levels.
- If design changes 1 and 2 are made and operating procedure A is modified, then the risk of operating the unloading facility can be reduced by a factor of 30.
- The major risk contributor in this process is failure of safety system C. Its failure contributes to 50% of the risk of this process.
- The estimated risk of a worker fatality during this new operation is 1000 times smaller than the risk from existing operations.

TABLE 7. Converting Absolute Risk Estimates to Relative Risk Estimates

System	Absolute Risk Estimate	Relative Risk Estimate
A	8×10^{-5}/y	1
B	2×10^{-5}/y	0.25
C	4×10^{-4}/y	5

along with design/operating cost figures, to rank these design options and ultimately select the best, most efficient process design.

Another way of normalizing absolute risk results is to use an existing risk estimate as the baseline case. For example, managers may need a quantitative comparison of the risk of a proposed new process to the risk of a current design. The results of a QRA performed on the earlier design are used to normalize the risk estimates for the new design. This method can also be used to compare the merits of different safety improvement recommendations for existing facilities. *However, the managers should be cautioned that unless the new study was performed under the same boundary conditions as the earlier study, the baseline results may not be appropriate for comparison purposes—different models, assumptions, and data may have been used in the earlier analysis, which would invalidate the comparison.*

Perhaps the easiest way to develop risk estimates for several design options is to pick a piece of input data common to all options and scale the input data for the designs relative to one of them. Consider, for example, three systems (A, B, and C) that each have different material handing requirements. System B will require twice as many material transfers as System A; however, the maximum amount of material that could be released from System B as a result of any one accident is one-third as much that could be released from System A. System C will require four times as many material transfers as System A, but the material involved is only half as toxic as the material in System A. Using material transfer frequencies of 1/week, 2/week, and 4/week for Systems A, B, and C, respectively, an analyst can then calculate accident sequence frequencies and consequences in a normal fashion. The result is a directly derived set of relative risk comparisons from which a decision to select the best design can be made. One advantage of this approach of scaling input data is that the analyst does not have to first calculate absolute risk estimates before normalizing them to arrive at the desired relative risk comparisons. A more universal approach is to select a constant divisor (e.g., $10,000/y) for all calculated risk results. The resulting risk index numbers can then be compared globally to show differences in design alternatives, processes, or facilities.

The use of relative results alone could encourage managers to make unnecessary improvements. Decision makers must use their judgment to make these decisions based on other information (e.g., qualitative results, codes and standards, industry practice, and intuition). They must determine whether to (1) explicitly choose a level of tolerable risk in using absolute risk estimates or (2) implicitly decide when sufficient changes have been made to a facility using relative results. In practice, using relative results is easier and preferable for some applications. Whenever possible you should charter QRA studies to provide relative risk results that support your particular needs (if you believe the problems associated with defending absolute estimates will detract appreciably from your ability to benefit from the study).

2.3. CRITERIA FOR ELECTING TO USE QRA

The decision whether to use QRA will be based on a number of factors, including the following:

- Do I have a reasonable expectation that the QRA can satisfy my needs?
- Is QRA the most efficient method?

To answer these questions you must consider details associated with your particular needs and activities of interest.

Figure 5 is an example of a decision tree you may find useful when considering QRA for particular process safety applications. The decision tree illustrates a flowchart of questions you can ask yourself (or others) to decide how far through the process of risk analysis to go to satisfy a need for increased risk understanding.

Step 1 considers all of the background information discussed in Section 2.1. If the information requirement is based on a regulatory concern or a special purpose need, then Steps 2 through 5 are bypassed and a QRA should be performed. If the information is needed for decision making, you must determine whether the significance of the decision warrants the expense of a QRA. If not, you may be able to use less resource-intensive qualitative approaches to satisfy your information requirements. Table 8 contains examples of typical conclusions reached from qualitative risk analysis results.

In Steps 2 through 5 of Figure 5 you will use subjective judgment to consider whether the situation involves major hazards, familiar processes, large consequence potential, or frequent accidents. The

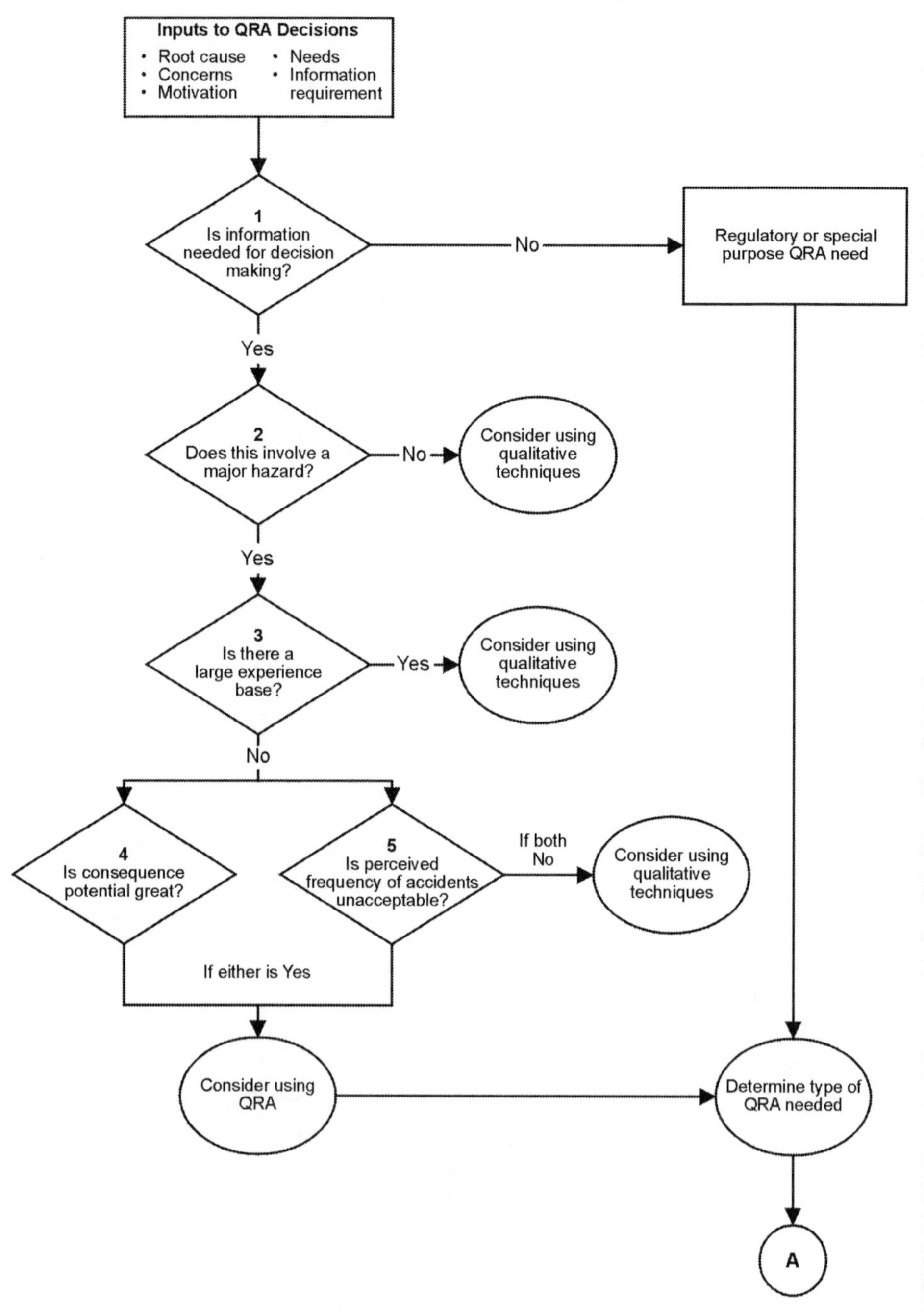

FIGURE 5. Decision criteria for selecting QRA.

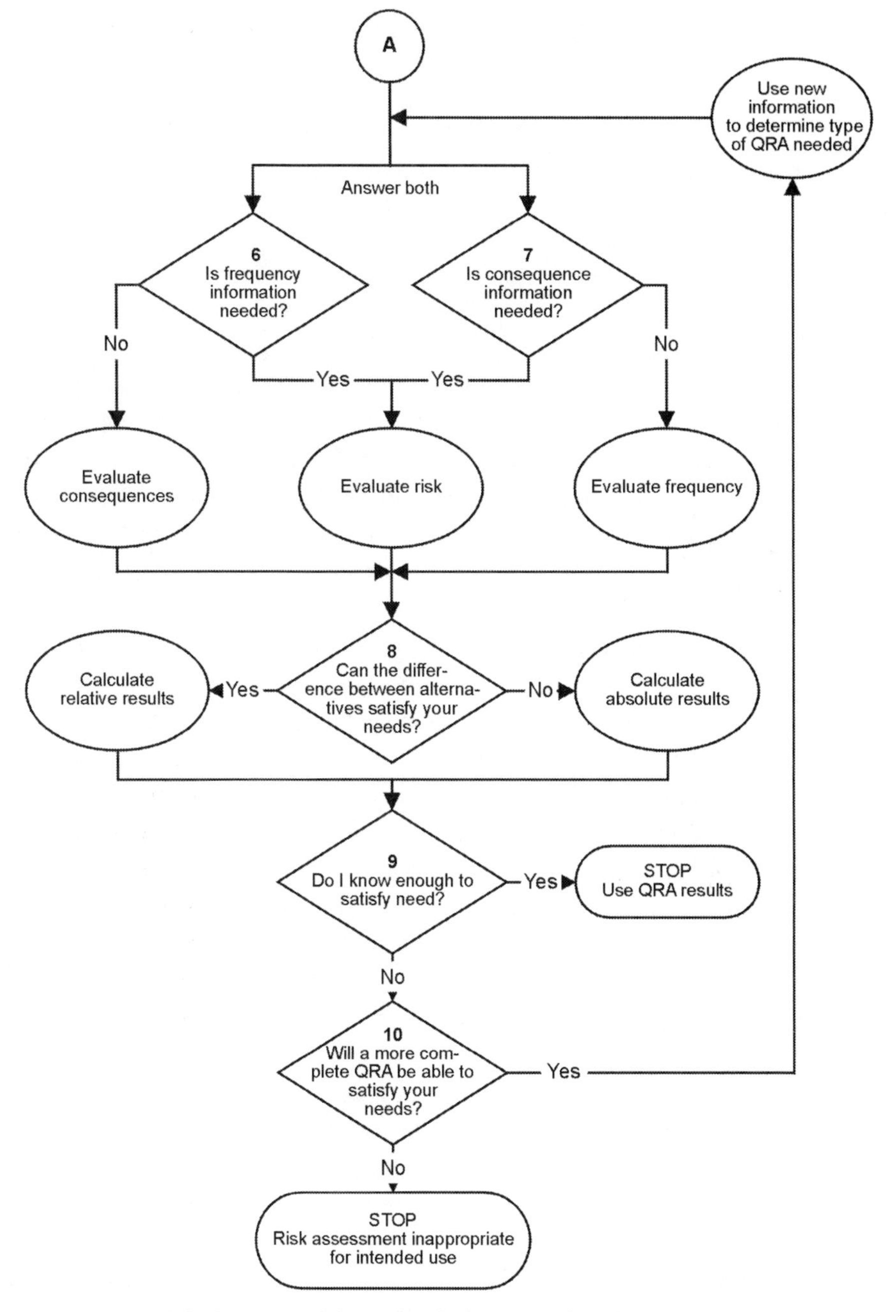

FIGURE 5. Decision criteria for selecting QRA (*cont.*).

TABLE 8. Examples of Possible Conclusions Using Qualitative Results

- There is/is not a significant hazard associated with this plant.
- There are few/many things that can go wrong and cause the accident of concern.
- The effects of a hypothetical accident are likely/unlikely to be bad.
- Implementing the following production capacity improvements will increase/decrease safety.

definition of major hazard (Step 2) may vary considerably from company to company, but managers should consider the inherent and intrinsic threat posed by the activity of interest (fire, explosion, toxic material release, etc.). A small hydrogen fire at a refinery pipe flange may be considered trivial, but that same fire in a semiconductor manufacturing clean room may be catastrophic. Even if the hazard potential is great, a company may have a large amount of relevant experience to base safety-related decisions upon, and QRA may not be required.

If sufficient experience does not exist, you should consider whether the consequence potential (Step 4) or the expected frequency of accidents (Step 5) is great. Consideration of consequence potential should include personnel exposure, public demographics, equipment density, and so forth in relation to the intrinsic hazard posed by the material of concern. In Step 5 you may perceive that the expected frequency of accidents alone is important enough to justify a QRA. However, even though your company may not have much relevant experience with the activity of interest, if the consequence potential of these accidents is not great, you may conclude that the expected frequency of the potential accidents is low enough for you to make your decisions comfortably using qualitative information alone.

Once a decision to use QRA has been made, you must decide whether frequency and/or consequence information is required (Steps 6 and 7). In some cases you may simply need frequency information to make your decision. For example, suppose you wish to evaluate the adequacy of operating procedures and safety systems associated with a chemical reactor. The main hazard of concern is that the reactor could experience a violent runaway exothermic reaction. You believe that you know enough about the severe consequences of a runaway and nothing more will be gained by quantifying the consequences of potential runaways. Instead, you decide to estimate the expected frequency of reactor upsets and safety system failures that could lead to reactor runaways. You use this estimate to identify weaknesses in the reactor operating procedures and protection system and to determine the most efficient ways to reduce the frequency, and therefore the risk, of reactor accidents.

In other cases the opposite may be true—you may decide it is more fruitful for you to base your decision on the results of a consequence analysis alone. For example, suppose you wish to evaluate and select the best combination of design and release mitigation features for a proposed facility for storing a highly toxic and reactive material. You may believe that your design team has already established the best engineering approach for preventing accidents. But, you are still concerned about the safety/health effects of a release and what emergency response capabilities you should establish. You have your QRA analysts quantify the possible effects of a release, assuming a worst-case release occurs, to provide you with information on which to base your selection of emergency response capabilities.

Whenever possible, relative comparisons of risk should be made (Step 8). Comparing relative risk estimates for alternative strategies avoids many of the problems associated with interpreting and defending absolute estimates. Table 9 contains examples of typical conclusions you can reach using relative risk estimates. In some cases, however, absolute estimates may be required to satisfy your needs. Table 10 contains a list of examples of typical conclusions possible using absolute risk estimates.

Once the QRA results are available, you must evaluate the information and determine whether it fully satisfies your needs (Step 9). If so, the results should be put into an appropriate format for communication to other parties (Section 4.3).

TABLE 9. Examples of Possible Conclusions Using Relative Risk Estimates

- Option A has lower risk than Option B.
- If A occurs, C is the most likely cause. Therefore, further risk-reduction efforts should be first directed toward C.
- If we change the system, the risk decreases/increases by a factor of X. We elect to change/not change the system because the cost is reasonable/excessive.

TABLE 10. Examples of Possible Conclusions Using Absolute Risk Estimates

- Option A is better than Option B. Both Options A and B are (not) acceptable.
- The risk of A is X.
- There is a 50% chance that event C will occur during the lifetime of the plant.
- We expect to lose Y dollars per year as a result of fire/explosion accidents in this process unit.
- The chance of severely injuring someone because of detonation accidents in this area is D per year.
- Changing A to reduce risk to a tolerable level will cost B dollars.

On rare occasions you may find that, because of things learned during the QRA or because of changing needs or assumptions, the information available from the QRA is not satisfactory. At this point you should carefully consider whether additional QRA will be of help (Step 10), and if you determine it will not, you should discontinue the QRA.

The strategy represented in Figure 5 should cover most applications. To be effective, individual managers will need to adapt this generic strategy to fit the needs of the company and the scope of their responsibilities.

3

MANAGEMENT USE OF QRA

A man is too apt to forget that in this world he cannot have everything. A choice is all that is left him.
H. MATTHEWS

Once you decide to use QRA to satisfy a particular need, you must devote attention to three key areas:

- Chartering the analysis
- Selecting appropriate techniques
- Understanding the assumptions and limitations

Some of these areas involve actions that primarily you, the ultimate user, must take (e.g., carefully defining written objectives for the QRA project team). Other areas involve decisions that you will influence, but that should be left to the team's discretion (e.g., selection of specific analytical techniques). Still other areas will require your careful interaction and negotiation with the QRA team to ensure that their final product meets your needs (e.g., defining analysis scope and available resources).

These areas are interrelated, and decisions about one affect the others. Also, decisions concerning these areas are not simply made once, never to be considered again. You should review each area periodically as intermediate results are developed to ensure that the QRA remains on track. Ignoring any of these areas diminishes the likelihood that your QRA objectives will be satisfied.

3.1. CHARTERING THE ANALYSIS

If a QRA is to efficiently satisfy your requirement, you must specifically define its charter for the QRA project team. Figure 6 contains the various

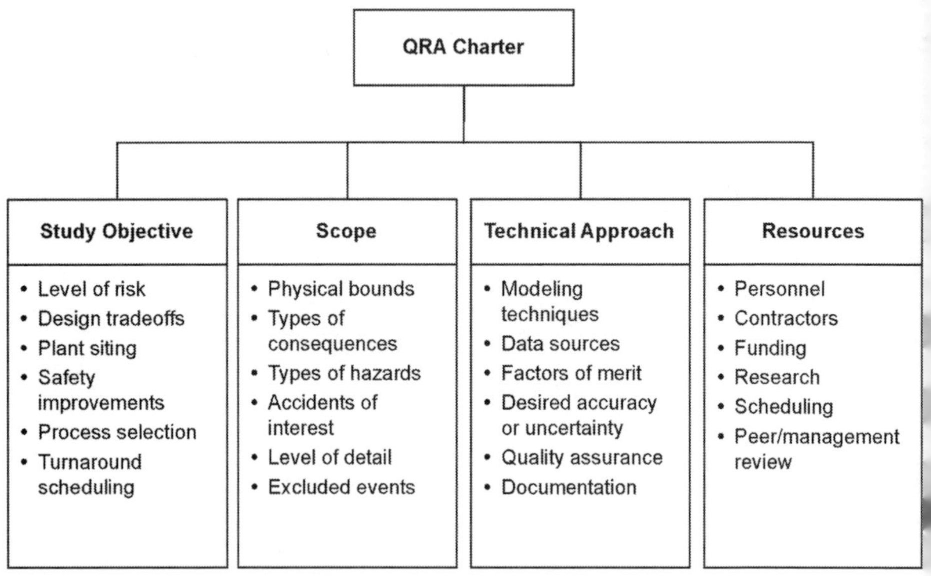

FIGURE 6. Elements of a QRA charter.

elements of a QRA charter. Defining these elements requires an understanding of the reason for the study, a description of the manager's needs, and an outline of the type of information required from the study. Sufficient flexibility must be built into the analysis scope, technical approach, schedule, and resources to accommodate later refinement of any undefined charter element(s) based on knowledge gained during the study. The QRA team must understand and support the analysis charter; otherwise a useless product may result.

3.1.1. Study Objective

An important and difficult task is concisely translating your requirements into study objectives. For example, if you need to decide between two methods of storing a hazardous chemical in a plant, the analysis objective should precisely define that what is needed is the relative difference between the methods, not the more general "I want to know the risk of these two storage methods." And asking your QRA team for more than is necessary to satisfy your particular need is counterproductive and may create unnecessary liabilities. For any QRA to efficiently produce the necessary types of results, you must clearly communicate your requirements through well-written objectives. Imposing vague requirements to "perform QRA" is not a workable strategy. Table 11 gives some examples of practical, achievable objectives for QRA.

3.1.2. Scope

Establishing the physical and analytical boundaries for a QRA is also a difficult task. Even though you will provide input, the scope definition will largely be made by the QRA project team. Defining the physical boundaries is relatively straightforward, but it does force the QRA team to explicitly identify and account for interfaces that may significantly affect the QRA results. For example, analysts often treat a connection to a power supply (e.g., a plug) or a feed source as a physical boundary; yet, loss of power or contamination of the feed must be considered in the QRA model.

Defining the analytical boundaries is a more subjective task. The first requirement is to identify the types of accidents to be considered. Once that is decided, selection of an appropriate level of detail is the scope element that is most crucial to performing an efficient QRA. You should encourage your QRA project team to use approximate data and gross levels of resolution during the early stages of the QRA. Once the project team determines the design areas that are the largest contributors to risk, it can selectively apply more detailed effort to specific issues as the analysis progresses. This strategy will help conserve analysis

TABLE 11. Examples of Typical QRA Objectives

- Determine if placing the process reactor in a containment cell will significantly reduce risk.
- Determine whether a catastrophic failure of the ammonia storage tank could cause irreversible health impacts in a nearby neighborhood.
- Identify the major risk contributors in a chemical unloading operation, and identify the best way to improve safety.
- Compare three process designs and rank them according to their risk to the community from both facility and transportation accidents.
- Investigate the potential for unconfined vapor cloud explosions resulting from accidents at the flammable storage tank area.
- Determine whether process improvements are needed to reduce the frequency (or consequences) of accidents.

resources by focusing resources only on areas important to developing improved risk understanding. You should review the boundary conditions and assumptions with the QRA team during the course of the study and revise them as more is learned about key sensitivities. *In the end, your ability to effectively use QRA estimates will largely be determined by your appreciation of important study assumptions and limitations resulting from scope definition.*

3.1.3. Technical Approach

The QRA project team can select the appropriate technical approach once you specify the study objectives, and together you can define the scope. A variety of modeling techniques and general data sources (discussed in Section 3.2) can be used to produce the desired results. Many computer programs are now available to aid in calculating risk estimates, and many automatically give more "answers" than you will need. The QRA team must take care to supply appropriate risk characteristics that satisfy your study objectives—and no more.

You should consider obtaining internal and external quality assurance reviews of the study (to ferret out errors in modeling, data, etc.). Independent peer reviews of the QRA results can be helpful by presenting alternate viewpoints, and you should include outside experts (either consultants or personnel from another plant) on the QRA review panel. You should also set up a mechanism wherein disputes between QRA team members (e.g., technical arguments about safety issues) can be voiced and reconciled. All of these factors play an essential role in producing a defendable, high-quality QRA. Once the QRA is complete, you must formally document your response to the project team's final report and any recommendations it contains.

3.1.4. Resources

Managers can use QRA to study small-scale, as well as large-scale, problems. For example, a QRA can be performed on a small part of a process, such as a storage vessel. Depending upon the study objectives, a complete QRA (both frequency and consequence estimates are made) could require as little as a few days to a few weeks of technical effort. On the other hand, a major study to identify the hazards associated with a large process unit (e.g., a unit with an associated capital investment of $50 million) may require 2 to 6 person-months of effort, and a complete QRA of that same unit may require 1 to 3 person-years of effort.

If a QRA is commissioned, you must adequately staff the QRA team if it is to successfully perform the work. An appropriate blend of engineering and scientific disciplines must be assigned to the project. If the study involves an existing facility, operating and maintenance personnel will play a crucial role in ensuring that the QRA models accurately represent the real system. In addition to the risk analyst(s), a typical team may also require assistance from a knowledgeable process engineer, a senior operator, a design engineer, an instrumentation engineer, a chemist, a metallurgist, a maintenance foreman, and/or an inspector. Unless your company has significant in-house QRA experience, you may be faced with selecting outside specialists to help perform the larger or more complex analyses. Depending on your QRA requirements, the team leader may require the assistance of specialists in modeling system failures and estimating their likelihood; specialists in modeling the release and dispersion of process materials; and specialists in modeling the effects of fires, explosions, and toxic material exposure. Even if outside experts (e.g., corporate specialists, contractors) are used extensively, you should require that your technical personnel be an integral part of the QRA team, and you should expect that they will spend a significant amount of time supporting the analysis work. No outsider understands your facility better than *your* personnel!

Adequate support from the facility staff is absolutely essential. The facility staff must help the analysis team gather pertinent documents (e.g., P&IDs, procedures, software descriptions, material inventories, meteorological data, population data) and must describe current operating and maintenance practices. The facility staff must then critique the logic model(s) and calculation(s) to ensure that the assumptions are correct and that the results seem reasonable. The facility staff should also be involved in developing any recommendations to reduce risk so they will fully understand the rationale behind all proposed improvements and can help ensure that the proposed improvements are feasible. Table 12 summarizes the types of facility resources and personnel needed for a typical QRA.

TABLE 12. Types of Facility Resources/Personnel

Engineering	The QRA team will need specific data on the design capabilities of the system being analyzed. For example, in a given scenario, how many coolers are required to remove the excess heat, how many pumps are required to deliver the minimum flow, or how many relief devices must operate to maintain safe pressure levels? In most QRAs, the primary resources required are process engineers and instrument/control engineers, but additional information may be required from structural engineers, electrical engineers, facility engineers, and other specialists. Expect to commit one full-time equivalent for the life of the project.
Scientists	The QRA team may need specific data on the kinetics of the system being analyzed. For example, what contaminants can trigger a runaway reaction, at what temperature does the substance decompose, and what by-products result when the material is released to the air? Chemists, physicists, researchers, and others may be needed to provide data. Expect to commit a few staff-weeks of effort during the project.
Operations	The QRA team will need specific data on how the system is actually operated. For example, are the bypass valves normally left open to increase throughput, what happens when the high level alarm sounds, or do operators bypass interlocks to continue production? Human actions/errors are usually dominant contributors to the real-world risks, and truthful data on actual process operations are vital to credible QRA results. Expect to commit one full-time equivalent for the life of the project.
Maintenance/ reliability engineering	The QRA team will need information about process configurations during maintenance activities and historic data on equipment performance. For example, how often are the instruments calibrated, does the same person calibrate redundant instruments, and how often do the pumps require repair? QRA results are always more accurate if based on actual plant data, and maintenance/testing policies can make an order of magnitude difference in the calculations. Expect to commit about 1 staff-week per month over the life of the project.
Emergency response personnel	If the QRA involves consequence estimation, the team will probably need data on the emergency response procedures and capabilities. For example, how quickly could the neighborhood be evacuated, what fraction of the release would be neutralized, and can firefighters extinguish a blaze in that location? Expect to commit a few staff-days of effort over the life of the project.
Safety/ industrial hygiene	If the QRA involves consequence estimation, the team will need data on the hazards of the materials that could be released. For example, at what concentration does this chemical cause irreversible damage if exposure time exceeds 10 minutes? Expect to commit a few staff-days of effort over the life of the project.

3.2. SELECTING QRA TECHNIQUES

Performing a QRA involves four steps:

- Hazard identification
- Consequence analysis
- Frequency analysis
- Risk evaluation and presentation

A multitude of analysis techniques and models have been developed to aid in performing these four steps (Figure 7). Many references exist for specific methods, and several recent publications give specific advice and "how to" details for the various techniques.[22–25] You will not have to select specific techniques—your QRA team will do that. But you must appreciate the types of results available from each class of techniques.

3.2.1. Hazard Identification

Hazard identification builds the foundation on which subsequent quantitative frequency and/or consequence estimates are made. Many companies have been using the hazard identification techniques listed in Figure 7 for

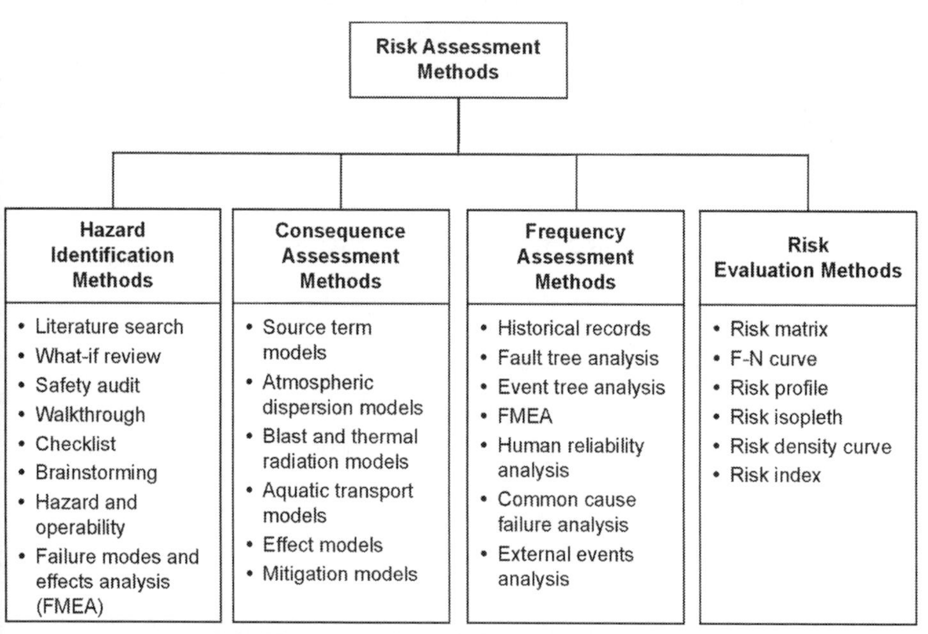

FIGURE 7. Overview of risk analysis methods.

years with great success. Generally, these methods yield a list of accident situations that could result in a variety of potential consequences. AIChE has published a book describing the most widely used hazard identification methods and the factors to consider when selecting one.[22]

The hazard identification step of the QRA typically requires the greatest involvement of plant personnel. For an existing process, only plant personnel know the status of process equipment and the current operating and maintenance practices. Excluding those personnel from the hazard identification step increases the chance of overlooking important potential hazards. For accurate results, the QRA team must have access to this information.

The cost of performing the hazard identification step depends on the size of the problem and the specific techniques used. Techniques such as brainstorming, what-if analyses, or checklists tend to be less expensive than other more structured methods. Hazard and operability (HAZOP) analyses and failure modes and effects analyses (FMEAs) involve many people and tend to be more expensive. But, you can have greater confidence in the exhaustiveness of HAZOP and FMEA techniques—their rigorous approach helps ensure completeness. However, no technique can guarantee that all hazards or potential accidents have been identified. Figure 8 is an example of the hazards identified in a HAZOP study. Hazard identification can require from 10% to 25% of the total effort in a QRA study.

3.2.2. Consequence Analysis

The consequence analysis step involves four activities:

- Characterizing the source of the release of material or energy associated with the hazard being analyzed
- Measuring (through costly experiments) or estimating (using models and correlations) the transport of the material and/or the propagation of the energy in the environment to a target of interest
- Identifying the effects of the propagation of the energy or material on the target of interest
- Quantifying the health, safety, environmental, or economic impacts on the target of interest

HAZOP Review of the Chlorine Railcar Unloading System
4 CHLORINE VAPORIZER

Drawings/Procedures - Drawing: 1

No.	Deviation	Causes	Consequences	Safeguards	Recommendations
4.4	Low temperature	Operator failing to open or inadvertently closing a valve in the steam system Operator setting steam pressure control valves incorrectly Steam pressure control valve failing to open or transferring closed Vaporizer tube fouling	Low temperature - chlorine gas supply line for bleaching line A (linked to 5.7)	Temperature indication and low temperature alarm in control room for the vaporizer that is indicated by the selector switch	30. Consider using a more direct control scheme for vaporizing the chlorine liquid (e.g., controlling the temperature at the discharge of the vaporizer by regulating steam flow)
4.5	High pressure	Low/no flow - chlorine gas supply line for bleaching line A (linked from 5.2) High temperature (linked from 4.3)	High pressure - chlorine railcar (linked to 1.5) Potential damage to the vaporizer if isolated from the relief valve on the chlorine railcar (linked to 4.9)	Local pressure indication	
4.6	Low pressure	Operator using the neutralization system eductor to remove chlorine from the vaporizer	Potential vaporizer collapse during chlorine evacuation through the eductor in the neutralization system, resulting in a small release of chlorine		5. Verify that all of the chlorine unloading and vaporizing equipment (particularly the chlorine railcars and vaporizers) can withstand the maximum vacuum created by the neutralization system. If the neutralization system can produce a vacuum capable of damaging equipment, consider modifying the neutralization system to minimize this potential

FIGURE 8. Example of a HAZOP table.

Many sophisticated models and correlations have been developed for consequence analysis.[23, 26–28] Millions of dollars have been spent researching the effects of exposure to toxic materials on the health of animals; the effects are extrapolated to predict effects on human health. A considerable empirical database exists on the effects of fires and explosions on structures and equipment. And large, sophisticated experiments are sometimes performed to validate computer algorithms for predicting the atmospheric dispersion of toxic materials. All of these resources can be used to help predict the consequences of accidents. But, you should only perform those consequence analysis steps needed to provide the information required for decision making.

The results from the consequence analysis step are estimates of the statistically expected exposure of the target population to the hazard of interest and the safety/health effects related to that level of exposure. For example:

- One hundred people will likely be exposed to air concentrations above the emergency response planning guidelines (e.g., ERPG-2, see Glossary).
- We expect 10 fatalities if this explosion occurs.
- If this event occurs, 1200 pounds of material is expected to be released to the environment.

The form of a consequence estimate is a direct function of the objectives and scope of the study. Consequences are usually stated in expected number of injuries or casualties or, in some cases, exposure to certain levels of energy or concentrations of substances. These estimates customarily use average meteorological conditions and population distribution, and may include mitigating factors such as evacuation and sheltering. In some cases simply assessing the quantity of material or energy released will provide an adequate basis for decision making. Figure 9 is an example of consequence analysis results from a typical QRA.

Like frequency estimates, consequence estimates can have very large uncertainties. Estimates that vary by orders of magnitude can result from (1) basic uncertainties in chemical/physical properties, (2) differences in average vs. time-dependent meteorological conditions, and/or (3) uncertainties in the release, dispersion, and effects models. Some

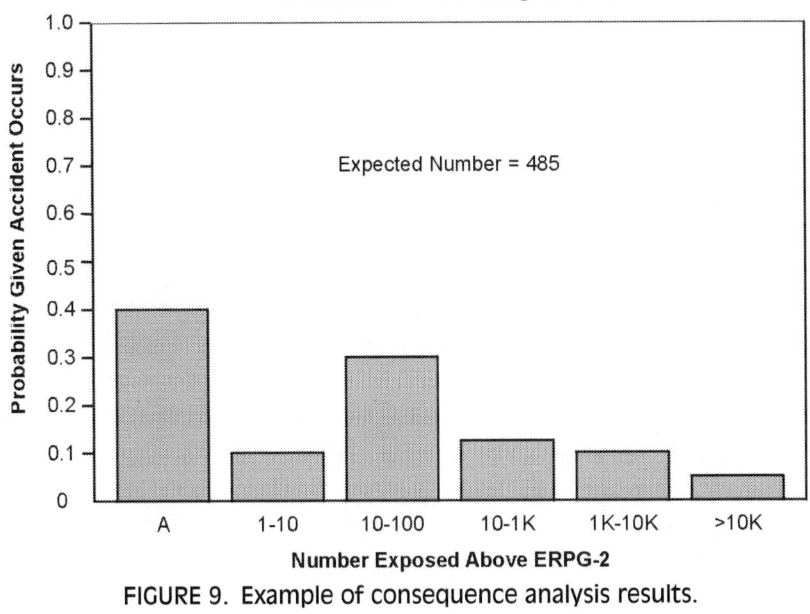

FIGURE 9. Example of consequence analysis results.

experts believe there is greater uncertainty in producing consequence estimates than in producing frequency estimates; others believe that the opposite is true. Either assertion is arguable and problematic.

In any case, like frequency analysis, examining the uncertainties and sensitivities of the results to changes in boundary conditions and assumptions provides greater perspective. The level of effort required for a consequence analysis will be a function of the number of different accident scenarios being analyzed; the number of effects the accident sequence produces; and the detail with which the release, dispersion, and effects on the targets of interest is estimated. The cost of the consequence analysis can typically be 25% to 50% of the total cost of a large QRA.

A typical phased QRA project to obtain consequence results is:

- *Consequence Phase 1:* **Identify Consequence Types and Screening Thresholds.** This activity is necessary to address Step 4 ("Is consequence potential great?") in Figure 5. Generally, consequence types for risk analyses are (1) employee safety, (2) public safety,

(3) environmental impact, (4) facility damage, and (5) production loss. The thresholds of concern will vary, but certainly consequence analyses will not be performed for accident scenarios that result in minor employee injuries (e.g., paper cuts).

- *Consequence Phase 2*: **Apply Qualitative Binning of Consequence Categories.** Define qualitative categories (e.g., negligible, marginal, critical, and catastrophic) for specific consequence types. The analysis team defines these categories, and the expected effects of the identified accident scenarios are placed in the appropriate bin.[12, 13] This consequence estimate, along with a frequency estimate (from the parallel frequency analysis path), will define the risk of the accident scenario.
- *Consequence Phase 3*: **Develop Detailed Quantitative Estimate of the Impacts of the Accident Scenarios.** Sometimes an accident scenario is not understood enough to make risk-based decisions without having a more quantitative estimation of the effects. Quantitative consequence analysis will vary according to the hazards of interest (e.g., toxic, flammable, or reactive materials), specific accident scenarios (e.g., releases, runaway reactions, fires, or explosions), and consequence type of interest (e.g., onsite impacts, offsite impacts, environmental releases). The general technique is to model release rates/quantities, dispersion of released materials, fires, and explosions, and then estimate the effects of these events on employees, the public, the facility, neighboring facilities, and the environment.

3.2.3. Frequency Analysis

The frequency analysis step involves estimating the likelihood of occurrence of each of the undesired situations defined in the hazard identification step. Sometimes you can do this through direct comparison with experience or extrapolation from historical accident data. While this method may be of great assistance in determining accident frequencies, most accidents analyzed by QRA are so rare that the frequencies must be synthesized using frequency estimation methods and models.

Synthesizing the frequencies of rare events involves (1) determining the important combinations of failures and circumstances that can cause the accidents of interest, (2) developing basic failure data from available industry or plant data, and (3) using appropriate probabilistic mathematics to determine the frequency estimates. Figures 10 and 11 illustrate simplified examples of the most frequently used models: events trees and fault trees. An event tree is often used to define all of the possible accident scenarios that could result from a particular upset initiating event.[22] Fault trees can be used to estimate the frequency or probability of individual events in an event tree.[22, 24, 29] Though limited, a few industry databases are available (see Table 2) from which to obtain generic data on component failure, and AIChE sponsored a project to develop a database specifically for the chemical industry.[14]

Layer of protection analysis (LOPA) is a simplified form of event tree analysis. Instead of analyzing all accident scenarios, LOPA selects a few specific scenarios as representative, or boundary, cases. LOPA uses order-of-magnitude estimates, rather than specific data, for the frequency of initiating events and for the probability the various layers of protection will fail on demand. In many cases, the simplified results of a LOPA provide sufficient input for deciding whether additional protection is necessary to reduce the likelihood of a given accident type. LOPAs typically require only a small fraction of the effort required for detailed event tree or fault tree analysis.

The frequency analysis step results in an estimate of an accident's statistically expected occurrence frequency. The estimates often take the

Event Tree Model

Initiating Event	Feed Shuts Off	Reactor Dump Works	Accident Sequence Number	Frequency (events/yr)	Consequence (impacts/event)
Loss of reactor cooling (2/yr)	0.9 (Success)		LOC-1	1.8	4-hour loss of production
	0.1 (Failure)	0.95	LOC-2	0.19	2-day loss of production
		0.05	LOC-3	0.01	Severe damage 3-month outage

FIGURE 10. Simplified example of event tree model.

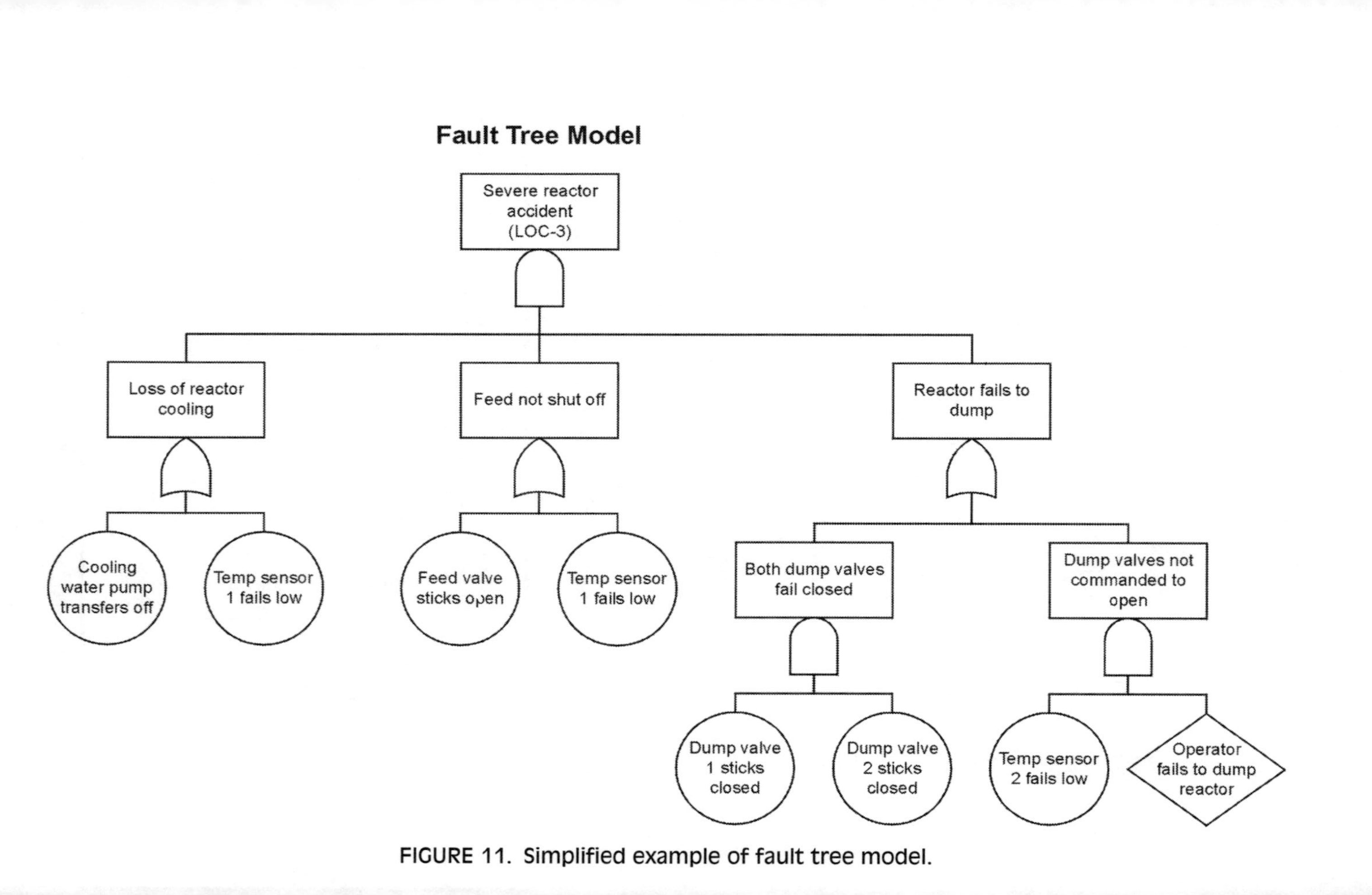

FIGURE 11. Simplified example of fault tree model.

form of very small numbers (e.g., 2×10^{-5} per year). Interpreting small numbers such as these is often a difficult task when evaluating risk-related results (Chapter 4).

If there is a lack of specific, appropriate data for a process facility, there can be considerable uncertainty in a frequency estimate like the one above. When study objectives require absolute risk estimates, it is customary for engineers to want to express their lack of confidence in an estimate by reporting a range estimate (e.g., 90% confidence limits of 8×10^{-6} per year to 1×10^{-4} per year) rather than a single-point estimate (e.g., 2×10^{-5} per year). For this reason alone it may be necessary for you to require that an uncertainty analysis be performed.

Many analysis methods and computer programs are available to simulate the variation in frequency analysis results that is due to data uncertainties. In addition, frequency analyses can be rerun under different sets of assumptions to determine the sensitivity of the results to important changes in boundary conditions. However, managers should be wary of the limitations of uncertainty analysis. Uncertainties result from a variety of causes. Uncertainty due to a lack of data is only one form, and often is not the most significant. Often the assumptions or the models dominate the uncertainty. A sensitivity analysis of the models, assumptions, and/or data often provides a better picture of the true uncertainty. (See Section 3.3, particularly Figure 15.) For most decisions, managers will have to rely on best estimates, compensating for any uncertainty with good judgment and intuition.

The level of effort required for a frequency analysis is a function of the complexity of the system or process being analyzed and the level of detail required to meet the analysis objectives. Frequency analysis can typically require 25% to 50% of the total effort in a large-scale QRA study. If an uncertainty analysis is performed, the effort required for the frequency analysis can be much greater.

A typical phased QRA project to obtain frequency results is:

- *Frequency Phase 1*: **Perform Qualitative Study, Typically Using HAZOP, FMEA, or What-if Analysis.** To perform a qualitative study you should first (1) define the consequences of interest, (2) identify the initiating events and accident scenarios that could lead to the consequences of interest, and (3) identify the equipment failure modes and human errors that could contribute to the accident scenarios. If this effort provides enough information to sufficiently answer the risk-based questions, then no additional phases are necessary. If the only need for the study is to identify ways to reduce risk and there is agreement that the risks should be reduced, a qualitative study alone is generally unacceptable.

- *Frequency Phase 2:* **Prepare Event Trees to Display Accident Scenarios of Interest.** An event tree is a graphical method of displaying the various accident scenarios that can result from an initiating event. An event tree starts with an initiating event that could result in a consequence of interest and postulates the success or failure of the safeguards designed to prevent or mitigate the accident. Review of qualitative event trees offers a forum for discussion of accident phenomenological information and could be sufficient to answer some risk-based questions.

- *Frequency Phase 3*: **Use Branch Point Estimates to Develop a Frequency Estimate for the Accident Scenarios.** The analysis team may choose to assign frequency values for initiating events and probability values for the branch points of the event trees without drawing fault tree models. These estimates are based on discussions with operating personnel, review of industrial equipment failure databases, and review of human reliability studies. This allows the team to provide initial estimates of scenario frequency and avoids the effort of the detailed analysis (Frequency Phase 4). In many cases, characterizing a few dominant accident scenarios in a layer of protection analysis will provide adequate frequency information.

- *Frequency Phase 4:* **Use Fault Trees to Solve the Initiating Event Frequency and Branch Point Probabilities.** Generally, fault trees are only needed if there is a great deal of dependency between the initiating event and the branch points or among the branch points themselves (e.g., if a pressure controller failure could cause a high pressure initiating event and also disable a high pressure shutdown interlock) or if there is a redundancy in the safeguard instrumentation (e.g., a 2-out-of-3 voting logic for a high pressure shutdown). If fault trees are not needed to answer the risk-based questions, then they should not be used.

3.2.4. Risk Evaluation and Presentation

Once frequency and consequence estimates are generated, the risk can be evaluated in many ways. It is essential that the large number of frequency/consequence estimates from a QRA be integrated into a presentation format that is easy to interpret and use. The presentation format you select will depend on the purpose of the QRA and the risk measure of interest.

Both societal (for large exposed populations) and individual (for single exposed persons) risk measures may be produced and presented. Individual risk is the likelihood that any one person will be injured within a given time period (typically per year or per working lifetime). Individual risk can be calculated for the maximally exposed employee, the average employee, the hypothetical person living at the plant fenceline, the nearest neighbor, the average member of the community, etc. Unfortunately, average individual risks can be manipulated to appear deceptively low if the analysts include a large number of unexposed or minimally exposed people in the average (e.g., including administrative, engineering, and management workers in the calculation of average employee risk). Societal risk is the grand total of all individual risks over a given time period (usually per year or per plant lifetime). Societal risk calculations are highly dependent on local population distributions and meteorological conditions, but are unaffected by attempts to include people who are not exposed to the hazard.

Both individual and societal risks may be presented on an absolute basis compared to a specific risk target or criterion. Or, they may be presented on a relative basis to avoid arguments regarding the adequacy of the absolute numbers while preserving the salient differences between alternatives. The end results of the risk presentation may be a single number (or a range of numbers if an uncertainty analysis was performed) or one or more graphs.

A common risk evaluation and presentation method is simply to multiply the frequency of each event by consequence of each event and then sum these products for all situations considered in the analysis. In insurance terms, this is the expected loss per year. The results of an uncertainty analysis, if performed, can be presented as a range defined by upper and lower confidence bounds that contain the best estimates. If the total risk represented by the best estimate or by the range estimate is below your threshold of concern (meets your risk goals), no additional information is necessary. But in other cases you will need additional risk information as a basis for decision making.

One danger in only using risk estimates presented as the product of frequency and consequence is losing your perspective on the types of accidents contributing to the risk. Are they high-frequency/low-consequence accidents that could be tolerable, or are they low-frequency/high-consequence accidents that would be catastrophic? Potentially severe accidents usually generate greater concern than smaller accidents, even though the risk (product) may be the same. To achieve a greater perspective, managers should request that their QRA team use one of several graphical devices to illustrate risk and the frequency/consequence relationship.

Figures 12 and 13 illustrate two of the more commonly used methods for displaying societal risk results: (1) an F-N curve and (2) a risk profile. The F-N curve plots the cumulative frequencies of events causing N or more impacts, with the number of impacts (N) shown on the horizontal axis. With the F-N curve you can easily see the expected frequency of accidents that could harm greater than a specified number of people. F-N curve plots are almost always presented on logarithmic scales because of

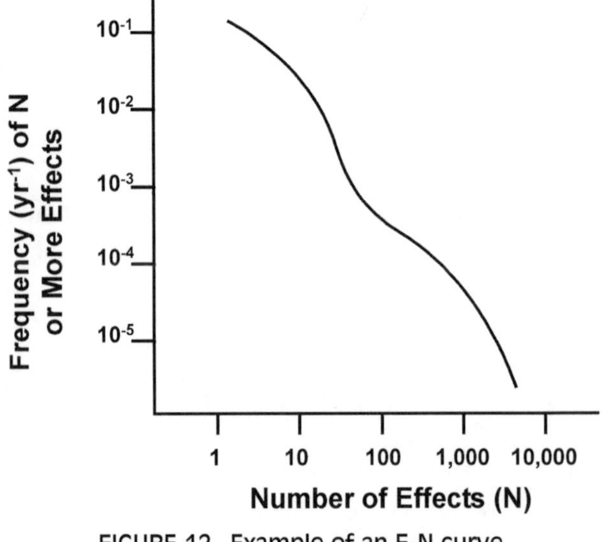

FIGURE 12. Example of an F-N curve.

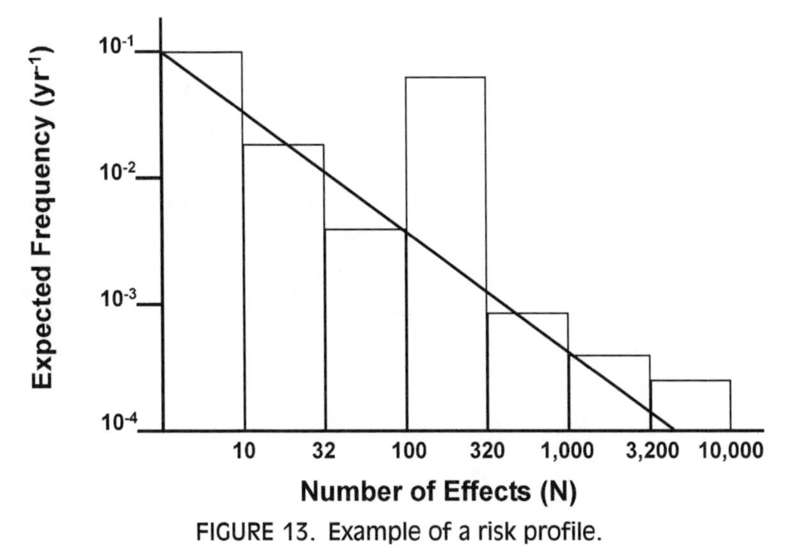

FIGURE 13. Example of a risk profile.

the wide range of data they contain. Users of F-N curves must be particularly careful to ensure their audience understands the graph's meaning.

While the F-N curve is a cumulative illustration, the risk profile shows the expected frequency of accidents of a particular category or level of consequence. The diagonal line is a line of constant risk defined such that the product of expected frequency and consequence is a constant at each point along the line.[30] As the consequences of accidents go up, the expected frequency should go down in order for the risk to remain constant. As the example illustrates, if a portion of the histogram "sticks its head up above the line" (i.e., a particular type of accident contributes more than its fair share of the risk), then that risk is inconsistent with the risk presented by other accident types. (Note: There is no requirement that you use a line of constant risk; other more appropriate risk criteria for your application can be easily defined and displayed on the graph.)

A method for graphically displaying individual risk results is use of the risk contour, or risk isopleth. If individual risk is defined as the likelihood of someone suffering a specified injury or loss, then individual risk can be calculated at particular geographic locations around the vicinity of a facility or operation. If the individual risk is calculated at many points surrounding the facility, then points of equal risk can be connected to create a risk contour map showing the geographic distribution of the individual risk. In Figure 14 you see various contours showing the probability of a particular impact on an individual located on the contour line.

The F-N curve, the risk profile, and the risk contour are the three most commonly used methods of graphically presenting risk results. Normally, you will elect to use more than one of these methods when evaluating risk estimates for decision making.

A valuable QRA result is the ***importance*** [26] of various components, human errors, and accident scenarios contributing to the total risk. The risk importance values highlight the major sources of risk and give the decision maker a clear target(s) for redesign or other loss prevention efforts. For example, two accident scenarios may contribute 90% of the total risk; once you realize that, it is obvious that you should first focus

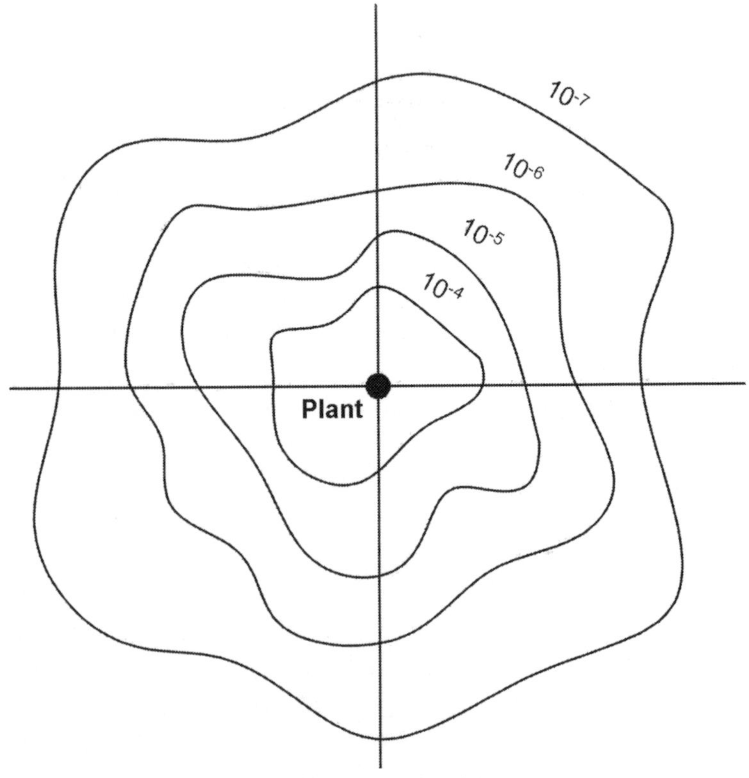

FIGURE 14. Example of a risk contour.

your loss prevention resources on reducing the potential for those accidents. In other cases all of the accident scenarios may have comparable risks, but failure of a process control computer is required for every scenario. The process computer will show up as the most risk-important component, and your loss prevention resources might best be spent in providing a backup computer. If you are using QRA to assist in decision making, you should always request risk importance results and seek cost-effective ways to eliminate or control the major risk contributors.

Another way to evaluate risks is to calculate the ***sensitivity*** of the total risk estimates to changes in assumptions, frequencies, or consequences. Risk analysts tend to be conservative in their assumptions and calculations, and the cumulative effect of this conservatism may be a substantial overestimation of risk. For example, always assuming that short-term exposure to chemical concentrations above some threshold limit value will cause serious injury may severely skew the calculated risks of health effects. If you do not understand the sensitivity of the risk results to this conservative assumption, you may misallocate your loss prevention resources or misinform your company or the public about the actual risk.

Risk sensitivity results are also very useful in identifying key elements in your existing loss prevention program. For example, suppose your fire protection system was assumed to have a very low probability of failure because you test it weekly. Fire protection failures may not show up as an important contributor to your total risk (because failure is so unlikely), but your total risk estimate may be extremely sensitive to any change in the probability of fire protection failures. Hence you should not divert resources away from testing the fire protection system unless the alternate use of funds will decrease risk more than the reduced testing will increase risk.

The work required to evaluate risk results will be a function of the objectives of the study. For relative risk studies, this evaluation is usually not very time-consuming. For absolute risk studies, in which many uncertainty and sensitivity cases may have been produced, the risk evaluation step may account for 10% to 35% of the total effort of a large-scale QRA. Chapter 4 discusses the problems associated with interpreting risk results.

3.3. UNDERSTANDING THE ASSUMPTIONS AND LIMITATIONS

Quantitative risk analysis is subject to several theoretical limitations.[31–33] Table 13 lists five of the most global limitations of QRA. Some of these may be relatively unimportant for a specific study, and others may be minimized through care in execution and by limiting one's expectations about the applicability of the results. However, you must respect these limitations when chartering a QRA study and when using the results for decision-making purposes.

3.3.1. Completeness

The hazard evaluation step is where the issue of completeness primarily arises. It is impossible for the QRA analyst to identify and model all of the things that can possibly go wrong, and it is impractical to evaluate the frequency and consequences of every identified event. But you can reasonably expect trained and experienced practitioners using systematic

TABLE 13. Classical Limitations of QRA

Issue	Description
Completeness	There can never be a guarantee that all accident situations, causes, and effects have been considered.
Model Validity	Probabilistic failure models cannot be verified. Physical phenomena are observed in experiments and used in model correlations, but models are, at best, approximations of specific accident conditions.
Accuracy/Uncertainty	The lack of specific data on component failure characteristics, chemical and physical properties, and phenomena severely limit accuracy and can produce large uncertainties.
Reproducibility	Various aspects of QRA are highly subjective—the results are very sensitive to the analyst's assumptions. The same problem, using identical data and models, may generate widely varying answers when analyzed by different experts.
Inscrutability	The inherent nature of QRA makes the results difficult to understand and use.

approaches and relevant experience data to identify the significant risk contributors. All QRAs involve screening out minor risks based on some qualitative judgment or criteria so the available resources can be focused on characterizing and understanding the major risks. However, there is no guarantee that all hazards have been identified or quantified, and this is an important limitation of risk analysis. Moreover, a QRA is a "snapshot in time" evaluation of a process. Any changes in the design or in the operating and maintenance procedures (however small) may have a significant impact on the QRA estimates.

3.3.2. Model Validity

The models you use to portray failures that lead to accidents, and the models you use to propagate their effects, are attempts to approximate reality. Models of accident sequences (although mathematically rigorous) cannot be demonstrated to be exact because you can never precisely identify all of the factors that contribute to an accident of interest. Likewise, most consequence models are at best correlations derived from limited experimental evidence. Even if the models are "validated" through field experiments for some specific situations, you can never validate them for all possibilities, and the question of model appropriateness will always exist.

3.3.3. Accuracy/Uncertainty

The accuracy of absolute risk results depends on (1) whether all the significant contributors to risk have been analyzed, (2) the realism of the mathematical models used to predict failure characteristics and accident phenomena, and (3) the statistical uncertainty associated with the various input data. The *achievable* accuracy of absolute risk results is very dependent on the type of hazard being analyzed. In studies where the dominant risk contributors can be calibrated with ample historical data (e.g., the risk of an engine failure causing an airplane crash), the uncertainty can be reduced to a few percent. However, many authors of published studies and other expert practitioners have recognized that uncertainties can be greater than 1 to 2 orders of magnitude in studies whose major contributors are rare, catastrophic events.

Some advocates of sophisticated data analysis and detailed uncertainty analysis contend that these approaches will engender greater confidence in the results. In fact, if the data are sparse, the models not extremely relevant, or the completeness of the study suspect, no amount of uncertainty analysis can help. As a practical matter, you will often base your decisions on best estimates—and your judgment.

3.3.4. Reproducibility

Probably the least appreciated weakness of QRA is that the results are difficult to duplicate by independent experts. Even with the variety of sophisticated tools available for use, QRA is still largely dependent on good engineering judgment. The subtle assumptions of analysts performing QRA studies can often be the driving force behind the results. Many times these assumptions are at best arguable, and at worst arbitrary.

A benchmark study examined the difficulty in reproducing QRA results.[33] Several expert teams were given identical systems to analyze using common techniques and a common database. The analysts were initially given total latitude concerning necessary assumptions, events to consider, data, and so forth. Figure 15 illustrates the results of the

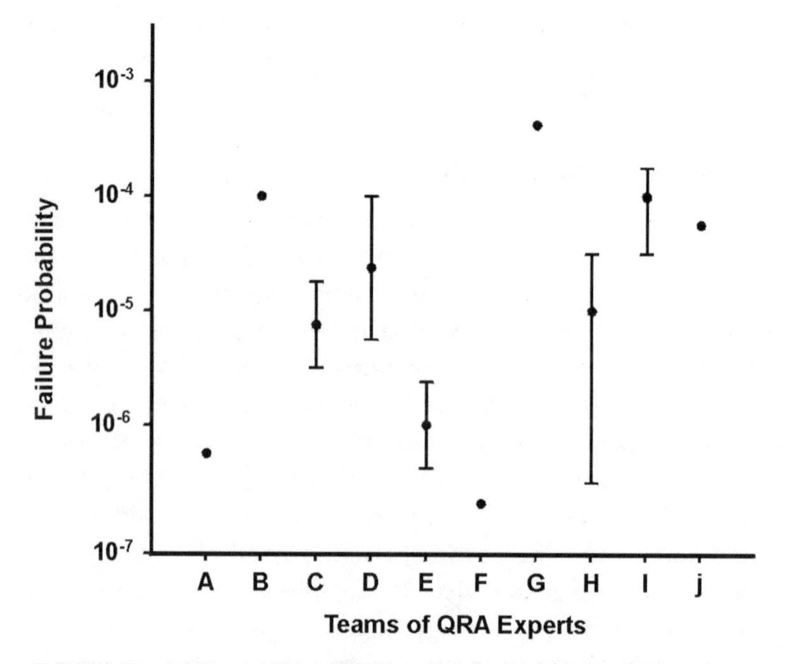

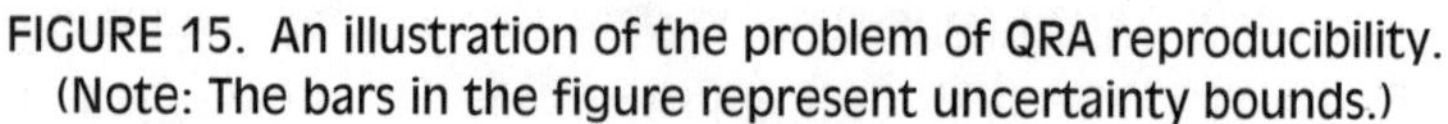
FIGURE 15. An illustration of the problem of QRA reproducibility. (Note: The bars in the figure represent uncertainty bounds.)

benchmark study. The best estimates of the factor of merit (in this case a probability of failure) ranged over several orders of magnitude—well beyond any of the uncertainty bounds calculated by some of the teams. Upon closer scrutiny, the researchers found that the different results arose from very basic (and very defendable, but different) assumptions used by the various analysis teams. Ultimately, when coached to use similar assumptions, the analysis teams' results converged within a reasonable range (i.e., within a factor of 5).

As a manager you must appreciate that the assumptions made during a QRA are as important as any quantitative result. And the decisions you make will be crucially tied to your appreciation of the limitations of such studies.

3.3.5. Inscrutability

QRA results can consist of many thousands of models, computer runs, calculations, and tables of numbers. Attempting to assimilate all of the details of an analysis is an overwhelming, tedious task. Combined with QRA analysts' tendencies to use large amounts of jargon, you will find yourself wondering what to do with it all. Using graphs and charts greatly improves the communication of risk results to decision makers and the public. You will have to depend on QRA experts to help you interpret the results until you gain greater QRA experience.

These limitations should not be reasons for rejecting the QRA approach. The solely retrospective approach of learning from experience is insufficient when the consequences of possibly rare accidents are severe. QRA provides a logical framework for examining hazards, using existing knowledge in an attempt to discover possible hazardous situations that may not have previously occurred. Simply because QRA is not perfect is no reason to completely reject using QRA to establish how severe accidents may occur or how significant these situations may be. Despite its flaws, QRA is sometimes the best tool for providing you with useful risk information.

4

USING QRA RESULTS

When you have to make a choice, and
don't make it, that in itself is a choice.
WILLIAM JAMES

Successful QRAs provide data and information that allow you to increase your wisdom and understanding of the risk of a particular activity. The usefulness of this information will ultimately be dictated by your ability to make sense of it. Moreover, the perspective resulting from such deliberations must be communicated to others (e.g., the public, regulators, senior management) if you are to effectively present cogent arguments using the risk estimates to support your decision-making purpose.

Any attempt to interpret QRA results must begin with a review of the analysis objective(s). If your objective was to identify the most important contributors to potential accidents, then the results may be completely unsuitable for presentation to zoning commissioners interested in the total risk of a toxic material release. *It is essential that QRA results be interpreted only in the context of the study objective(s).*

Four essential areas largely determine your success in capitalizing on high-quality QRA results:

- Presenting the results in perspective
- Recognizing the factors that influence perceptions of the meaning of the results
- Credibly communicating risk information in the public arena
- Avoiding common pitfalls in using the results for making the "right" decision

It is often helpful to talk to the QRA team members to determine their personal impressions and conclusions about the study. Often a great benefit of a QRA is the insight the analysts gain from having gone through this exercise. The more you can absorb these insights, the better able you will be to confidently interpret and use the results in making decisions.

4.1. COMPARATIVE METHODS FOR ESTABLISHING PERSPECTIVE

Quantitative risk analysis is a forecast concerning the degree of belief associated with the occurrence of future events. It normally focuses on those classes of events that are rarely expected to occur at a facility. However, because the potential consequences of such events may be so great, the possibility that the events could occur at all gives rise to concern. When a QRA generates results that reflect a very small likelihood of an event and confirm the suspicion that the event could have a severe impact, these questions inevitably arise: What does it all mean? What should I do about it?

The problems with interpreting absolute risk estimates usually outweigh the difficulties with understanding relative risk estimates. Use of absolute risk results requires a mature and cautious attitude toward the accuracy of the estimates. Studies designed to produce relative estimates are mandated to help answer the question, is Option A significantly better than Option B? With these results you usually need to become comfortable with only the robustness or accuracy of the comparison; deciding to go with the safer option is perfunctory. Only when cost becomes a significant factor (if B costs much more than A) does the management decision become more difficult. If the decision is whether to go beyond generally accepted minimum safety standards, managers must use their judgment to answer the question, are there other ways to spend these resources in other areas of the company that would provide greater risk reduction?

Absolute risk estimates can be difficult to use when there is no apparent human experience against which to calibrate them. By definition, there never exists enough experience about catastrophic rare events (fortunately) with which to calibrate the thinking about their significance. If there were enough data, you would not have elected to do the QRA in the first place. So, now that you have a "bottom line" estimate of risk, how do you figure out how accurate it is, whether it is tolerable, and what to do?

Consider the following example in which the worker risk from a catastrophic accident has been calculated to be 2×10^{-4} fatalities per year. It is possible to interpret this number in many ways, but one of the most common ways is the following: there is one chance in 5000 per year that a worker will be fatally injured at the plant. However, you should be cautious when interpreting single risk estimates that are the sums of products of frequency and consequence of many accidents. The way you believe (and act) may be affected by the frequency/consequence profile that the number represents (see Sections 3.2.4 and 4.2.5.) That is, your reaction to an accident that occurs once every 100 years and kills 1 person (Risk = 10^{-2} fatalities per year) and your reaction to an accident that occurs once every 10,000 years and kills 100 people (Risk = 10^{-2} fatalities per year) are likely to be very different.

There are several widely used approaches for developing perspective about the significance of absolute risk estimates (Figure 16).[34–37] The first approach is to compare the risk estimates to historical experience within your company, looking for similar events. Most companies have safety and loss recordkeeping programs that date back many years. But if directly related data are sparse, you may widen your comparison to extrapolate from near-miss incidents that could have caused the event of interest. You will not, however, frequently find solace from the company data—or even comparable industry data.

Another approach is to use government and private mortality and injury statistics. Calculated absolute risk estimates (the probability per year of a worker being injured or killed) can be compared to those de facto worker risk standards. For example, in the United Kingdom, industry and government alike are using the fatal accident rate (FAR, see Glos-

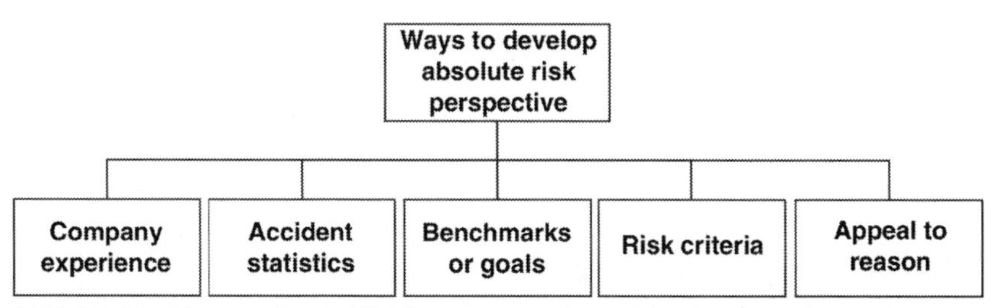

FIGURE 16. Means of establishing perspective with absolute risk estimates.

sary) as a standard for establishing the tolerability of the estimated risk for industrial plants.

If the probability of worker injury or death because of participation in a given work-related activity can be shown to be much less than the risk of injury or death associated with presently accepted activities under very similar circumstances (e.g., the same type of hazard), then you may feel more comfortable about accepting the status quo. Table 14 illustrates the types of public mortality data available for such comparisons.[38] In the previous example, where the worker risk was calculated as 2×10^{-4} fatalities

TABLE 14. Example of Mortality Statistics

Hazard	Total Number of Deaths in 1994[38]	Individual Chance of Death per Year[a]
Heart disease	732,400	0.0028
Cancer	534,300	0.0020
All accidents	91,400	0.00035
Motor vehicles	42,524	0.00016
Homicides	21,600[b]	0.000083
Falls	13,450	0.000052
Poisoning	8,994	0.000035
Work accidents	5,000	0.000019
Fires, burns	3,986	0.000015
Drowning	3,404	0.000013
Suffocation, ingested objects	3,065	0.000012
Firearms accidents	1,356	0.000005
Air transport accidents	1,075	0.000004
Pleasure boating	784	0.000003
Water transport	723	0.000003
Railway accidents	635	0.000002
Floods	72	<0.000001
Tornadoes	69	<0.000001
Hurricanes	38	<0.000001

[a]U.S. population in 1994 equal to 260,682,000.

[b]1995 statistic.

per year, the risk is comparable to the risk of dying in a motor vehicle accident.

Another way of interpreting absolute risk estimates is through the use of benchmarks or goals. Consider a company that operates 50 chemical process facilities. It is determined (through other, purely qualitative means) that Plant A has exhibited acceptable safety performance over the years. A QRA is performed on Plant A, and the absolute estimates are established as calibration points, or benchmarks, for the rest of the firm's facilities. Over the years, QRAs are performed on other facilities to aid in making decisions about safety maintenance and improvement. As these studies are completed, the results are carefully scrutinized against the benchmark facility. The frequency/consequence estimates are not the only results compared—the lists of major risk contributors, the statistical risk importance of safety systems, and other types of QRA results are also compared. As more and more facility results are accumulated, resources are allocated to any plant areas that are out of line with respect to the benchmark facility.

Having a numeric criterion for tolerable risk would be everyone's choice when making decisions using absolute risk estimates. Unfortunately, no universally accepted or mandated criterion exists. Nevertheless, when attempting to establish risk guidelines satisfying the requirements described, an organization has a number of resources available.[39–42] Some particularly valuable sources of information include:

- **Previous CPQRA results.** It is best to have some experience in doing chemical process quantitative risk analysis (CPQRA) studies before considering the establishment of quantitative risk guidelines. This is important in defining the organization's quantitative risk analysis "culture." A database of CPQRA results can be created, and the proposed guidelines compared to the studies already completed. The application of the proposed guidelines should, in general, make sense. Because decisions based on prior studies will have been made in the absence of quantitative guidelines, a few exceptions to decisions that would have been made based on the guidelines can be expected. A large number of past decisions that are inconsistent with the proposed new guidelines may suggest a problem, and the inconsistencies should be investigated.

- **Risk guidelines published by many organizations and governments.** While these guidelines have been developed in the context of a specific risk analysis methodology and culture, they do provide general guidance. Some examples of risk guidelines have been published by Ale,[43, 44] Goyal,[45] Health and Safety Executive,[46] Helmers and Schaller,[47] Pasman,[48] Renshaw,[49] and Whittle.[50] Pikaar and Seaman[51] have summarized the application of risk guidelines by both governments and companies in Europe and the United States.
- **Safety records for both the organization considering the establishment of guidelines, and the chemical industry in general.**[52, 53] A proposed risk guideline should ensure that facilities tested against the guideline and passing will be at least as safe as the organization's past safety record. Furthermore, the risk guideline should provide a driving force for continuing improvement of that safety record. The Center for Chemical Process Safety's (CCPS's) *Tools for Making Acute Risk Decisions with Chemical Process Safety Applications*[54] discusses the use of decision tools, including risk guidelines, in decision making.
- **Government and regulatory decisions.** Sometimes these decisions are based on some type of quantitative risk analysis, and they provide some guidance on society's expectations with regard to risk management. In some cases these decisions will also include some kind of cost-benefit analysis. The current political climate in the United States may encourage more extensive use of risk analysis in the establishment of future regulations.
- **General societal risk data.** It is important to consider the context of societal risk data. Some particularly important factors include whether or not the risk is voluntary, and whether persons exposed to the risk derive any benefit from the activity that generates the risk. Covello, Sandman, and Slovic, [55] Slovic,[56] and Wilson and Crouch[57] provide examples of general societal risk data and discuss risk comparison and perception.

However, because of the diversity of short-term and long-term hazards to the public and workers in the CPI, no single criterion will ever meet everyone's needs. Even if such a criterion could be developed, there would still be controversy over its use. Absolute risk estimates are not

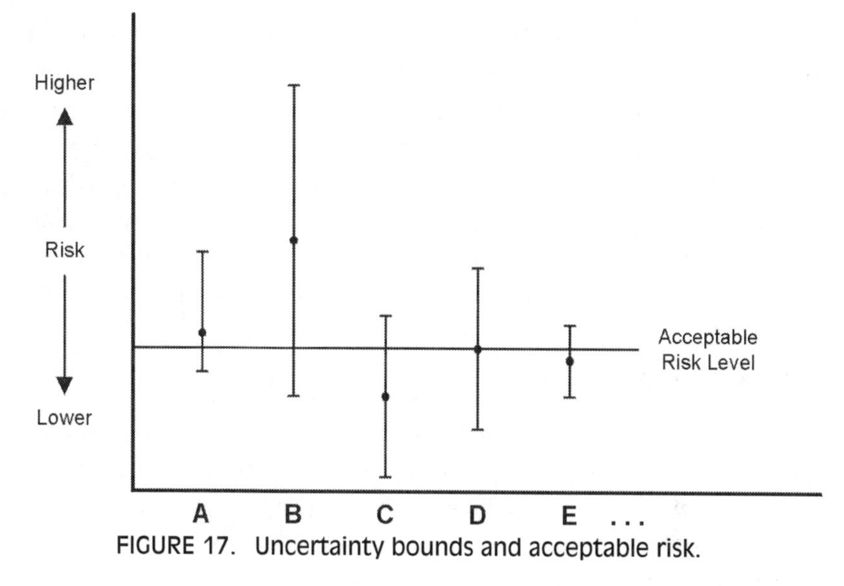

FIGURE 17. Uncertainty bounds and acceptable risk.

point values; they are ranges with uncertainty bounds and probability distributions associated with them. As illustrated in Figure 17, is a risk unacceptable if any portion of its uncertainty bound falls above the tolerable risk criterion? And how much confidence should one have that the true risk falls within the uncertainty bounds? Should uncertainty bounds encompass one standard deviation? Three? Ten?

The last method is simply an appeal to reason. If a QRA indicates that the risk of a member of the public dying because of an industrial activity is very low (e.g., less than one chance in some very large number), then the risk is negligible in comparison to other imposed risks commonly tolerated by our society (e.g., having an airliner crash into your home). However, such comparisons are often misleading because the risk per year does not necessarily reflect the risk per activity or the risk per hour of exposure.

4.2. FACTORS INFLUENCING RISK PERCEPTION

The effective use of risk results demands not only selecting appropriate means of establishing the credibility of the results, but also considering the audience who is or will inevitably become aware of or review those results. Risk perception has become a buzz topic over the past few years.[58–60] Its importance is universally accepted because of the tacit recognition that "it doesn't matter what the 'real' risk is, it's what people think the risk is." Risk communication research has found that many attributes can significantly affect the way people perceive risk. As a manager who uses QRA results, you must be cognizant of these influences on yourself and on others who are affected by your safety-related decisions. Table 15 outlines some of the more important perception issues.

4.2.1. Type of Hazard

The public's idea of what is most risky usually differs widely from the facts. Much of this stems from disproportionate media coverage of "newsworthy" hazards that are rare or unusual. This distorts many people's perspective on the relative magnitude of risks. Thus, some parents may not have their children vaccinated, fearing a rare reaction to the vaccine more than the overwhelmingly greater risk of dying from the disease.

The way a hazard manifests itself as a threat to an individual affects how that person feels about the risk. For example, the hazards of nuclear power are viewed as much worse than the prospect of being killed as a pedestrian, yet data show the risk of the latter is much greater than that of the former.

TABLE 15. Issues Affecting Perception of Risk

- Hazard type and effect
- Voluntary versus involuntary
- Societal versus individual
- Public versus employee
- High consequence/low frequency versus low consequence/high frequency
- Acute versus latent effects
- Distribution of benefits versus risk
- Familiarity
- Controllability
- Age of exposed population

4.2.2. Voluntary versus Involuntary

People will tolerate a greater level of risk if the threat is one they specifically have chosen to accept (mountain climbing, flying, etc.). Individuals reject comparable risk if the risks are imposed upon them (e.g., a landfill springing up in a hitherto vacant lot beside a house).

4.2.3. Societal versus Individual

Societal and individual risks are different presentations of the same underlying combinations of accident frequency and consequence estimates. However, they address the issues of risk to groups of people rather than to specified individuals. People are more willing to tolerate risks that confer threats to individuals or small groups (e.g., workers in a chemical plant). People tend to reject comparable risks that threaten large groups or society as a whole (e.g., the existence of an incinerator in the community). Both societal and individual risk measures are important in assessing the benefits of risk-reduction options or in judging the acceptability of a facility in absolute terms.

4.2.4. Public versus Employee

Sometimes people view higher levels of worker risk as being more tolerable than comparable levels of public risk. This is partially because the worker has voluntarily accepted the risk and is receiving direct benefits from the acceptance of that risk.

4.2.5. High Consequence/Low Frequency versus Low Consequence/High Frequency

Consider an economic risk example. Accident A for a plant has a frequency of once every 2 years and a consequence of $100,000, yielding a risk of $50,000 per year. Accident B in the same plant has a frequency of once every 10,000 years but a consequence of $500 million, yielding an equivalent risk. Managers typically react to these differences by giving more attention to the higher consequence event because, if it were to occur, it might mean the company's going out of business. Hence managers often set lower thresholds for accepting the risks of high-consequence/low-frequency events than for low-consequence/high-frequency events.[61] In the Netherlands, this aversion is incorporated in regulations that require the likelihood of an accident to drop two orders of magnitude for every order of magnitude increase in severity.

4.2.6. Acute versus Latent Effects

Most people will tolerate greater risk from activities when the threat to life is offset in time from when the risk (and the benefit) is originally accepted. For example, people may feel worse (and usually accept less risk) about a threat of immediate harm (e.g., the blast wave from an explosion) than a threat of latent harm (e.g., an increase in the chance of getting a fatal disease following a 20-year exposure to a hazardous material, like asbestos), even though the risks may be equivalent.

4.2.7. Familiarity

Individuals tend to acclimate themselves and their concerns (sometimes to their detriment) about the risk of a given activity if they have a large amount of personal experience in dealing with a well-known hazard. For example, an individual may accept the risk of driving a car on a busy highway but reject the much lower risk of flying in a commercial airliner.

4.2.8. Controllability

People are more comfortable when they are in control. Individuals tend to accept greater risk when they feel as though their actions can directly influence the possibility of experiencing an adverse effect from participation in a particular activity. For example, an automobile trip is viewed as less risky by the driver than by the passenger.

4.2.9. Age of Exposed Population

People are less willing to threaten the safety of younger people. School-age youngsters and babies are particularly important because they are viewed as the endowment of our future. Older people are also a special concern if they are not able to protect themselves (e.g., evacuate the area) without assistance.

4.2.10. Distribution of Risk and Benefit

People are more willing to accept risks from which they will receive a direct, tangible benefit.[62] A one-company town will likely have widespread community support for the company and accept the risks of its business—it directly or indirectly provides the livelihood for most families in the community. This may not be the case in an area having a broad-based manufacturing and service economy. Here the relatively small public benefit from a new plant may be outweighed by the public's perception of the plant's risk. People are unwilling to tolerate a given level of risk unless there is a direct benefit to themselves.

4.3. COMMUNICATING RISK

Sometimes the results of QRA will be used in the public arena, and communicating to the public about the risks of exposure to chemicals is difficult. You must be sensitive to the feelings of a public that is generally suspicious of industry and ignorant of science. As the *source* of risk information, it is your responsibility to communicate a *message* through a *channel* (meeting, newsletter, videotape, public service announcement, etc.) that the *receiver* (citizens, government officials, emergency responders, media, etc.) understands. Communication can be rewarding for *source* and *receiver* alike if *The Seven Cardinal Rules of Risk Communication*[63] are followed. (More resources on risk communication are listed in "Suggested Additional Reading.")

4.3.1. Accept and Involve the Public as a Legitimate Partner

A basic tenet of risk communication is that people have a right to participate in decisions that affect their lives. The goal of risk communication should be to inform the community about the risks and potential health effects of your activities and to involve the public in developing solutions to any related problems.

4.3.2. Plan Carefully and Evaluate Your Efforts

Risk communication will be successful only if it is carefully planned. Establish risk communication objectives, such as providing information to the public and motivating individuals to act. Evaluate your information and know its strengths and weaknesses. Aim your messages at your specific audience. There is no such entity as "the public." Instead, there are many publics, each with its own interests, needs, concerns, priorities, preferences, and organizations. Whenever possible, pretest your messages and, after each presentation, analyze how you can improve the next one.

4.3.3. Listen to People's Specific Concerns

If you do not listen to people, you cannot expect them to listen to you. Communication is a two-way activity. Do not make assumptions about what people know, think, or want done about risks. Take the time to find out what people are thinking. Often, people are more concerned about issues such as trust, credibility, competence, control, voluntariness, fairness, and compassion than about mortality statistics and the details of QRA. Use techniques such as interviews, focus groups, and surveys to gauge what people are thinking.

4.3.4. Be Honest, Frank, and Open

In communicating risk information, trust and credibility are imperative. If you do not know an answer, say so, then get back to those people when you do have an answer. Discuss data uncertainties, strengths, and weaknesses, including ones identified by other credible sources. Identify worst-case estimates as such, and cite ranges of risk estimates when appropriate.

4.3.5. Coordinate and Collaborate with Other Credible Sources

Devote time and resources to building bridges with other organizations. Use credible and authoritative intermediaries. Consult with others to determine who is best able to answer questions about risk. Few things make risk communication more difficult than conflicts or public disagreements with other credible sources.

4.3.6. Meet the Needs of the Media

The media are prime *channels* of information on risks, playing critical roles in setting agendas for public debate and determining the outcomes of those debates. Be open and accessible to reports. Respect the deadlines of reporters and provide risk information tailored to the needs of each type of media. Try to establish long-term relationships of trust with editors and reporters in your community.

4.3.7. Speak Clearly and with Compassion

Technical language and jargon are useful as professional shorthand, but they are barriers to successful communication with the public. Use simple, nontechnical language and vivid, concrete images that communicate on a personal level. Avoid distant, abstract, unfeeling language about deaths, injuries, and illnesses.

4.4. PITFALLS IN USING QRA RESULTS

There are a variety of things that can go wrong when using QRA. Recognizing these potential problems up front will enable you to charter and use QRA without incurring unnecessary expense or making a wrong decision based on inaccurate results. Table 16 lists a few of the more important situations that managers should avoid when using QRA.

TABLE 16. Typical Pitfalls in Using QRA

- Inadequately defining analysis scope and objectives
- Using QRA in situations where qualitative approaches would suffice
- Overworking the problem: analyzing more cases and using more complicated models than required to produce the necessary information for a decision
- Dictating that inappropriate analysis techniques be used
- Using inexperienced or incompetent practitioners
- Choosing absolute results when relative results would suffice
- Selecting an incorrect risk characteristic as a factor of merit
- Not providing sufficient resources (e.g., time, money, process experts)
- Having unrealistic expectations
- Being overly conservative
- Failing to acknowledge the importance of the analysis assumptions and limitations
- Modeling the design, not the reality, of plant operations
- Quantifying only the known hazards of a new or unproven technology

5

CONCLUSIONS

Don't find fault, find a remedy.
HENRY FORD

Quantitative risk analysis is an important tool for the CPI. In selected cases it can complement (not replace) other historically successful methods for safety assurance, loss prevention, and environmental control. QRA is a new, evolving technology, still more of an art than a science, that will never ***make*** a decision for you—it can only help increase the information base from which you will decide what to do. More conventional process safety management practices, such as good design standards, proper construction, accurate procedures, thorough training, and sound management judgment, will continue to form the foundation for a safe and productive chemical industry.

In the past, qualitative approaches for hazard evaluation and risk analysis have been able to satisfy the majority of decision makers' needs. In the future, there will be an increasing motivation to use QRA. For the special situations that appear to demand quantitative support for safety-related decisions, QRA can be effective in increasing the manager's understanding of the level of risk associated with a company activity. Whenever possible, decision makers should design QRA studies to produce relative results that support their information requirements. QRA studies used in this way are not subject to nearly as many of the "numbers" problems and limitations to which absolute risk studies are subject, and the results are less likely to be misused.

When managers are faced with the necessity of using QRA results on an absolute basis, they must respect the potentially large uncertainties associated with the numbers and use prudent and conservative interpretations of these results for their decisions. Absolute risk estimates in these cases must be viewed with caution and carefully scrutinized to learn what is behind the numbers rather than accepting the numbers at their face value.

Whenever the commitment to perform a risk analysis is made (especially for a quantitative analysis), managers should also recognize the implied commitment they make to take action based on the analysis results. If a QRA study results in recommendations for improving a process, selecting a plant site, and so forth, managers must be cognizant of the necessity to document the decision-making process, using the risk results to act on the recommendations of the study. It is imperative that managers recognize the potential legal implications of a situation in which a company, having performed a risk study prior to an incident, failed to respond to the recommendations from the study, neither implementing the risk-reduction alternatives nor justifying why they are unnecessary.

Because QRA reports are legally sensitive documents, the wise manager will involve legal counsel in the review of any work products. In particular, the QRA report should be as factual as possible and refrain from unnecessary opinions, speculation, and use of inflammatory language. When comparing risks to a "standard" or "guideline," the report should be clear about whether the system being reviewed is required to meet or exceed the point of comparison. The report should preserve management's options for addressing risks and avoid using words like "must," "shall," or "necessary." Also, technical reports should not draw legal conclusions or compromise legal protections available to the company.

When used judiciously, the advantages of QRA can outweigh the associated problems and costs. Companies that prudently commission QRAs and conscientiously act on the resulting recommendations are better off for two reasons: (1) they have a better base of information to make decisions and (2) their judicious use of QRA technology represents another demonstration of responsible concern for the health and safety of workers and the public. However, companies should resist the indiscriminate use of QRA as a means to solve all problems since this strategy could waste safety improvement resources, diverting attention from other essential safety activities. Once executives are able to interpret and use QRA results, they will appreciate that the quality of their decisions rests largely on their ability to understand the salient analysis assumptions and the limitations of the results.

REFERENCES

1. Vernon L. Grose, *Managing Risk—Systematic Loss Prevention for Executives*, Prentice Hall, Englewood Cliffs, NJ, 1987.
2. George L. Head, *The Risk Management Process*, Risk and Insurance Management Society, Inc., New York, NY, 1978.
3. William C. Wood, *Nuclear Safety: Risks and Regulations*, American Enterprise Institute for Public Policy, Washington, DC, 1983.
4. Chauncey Starr, "Social Benefit versus Technological Risk," *Science*, Vol. 165, September 19, 1969, pp. 1232–1238.
5. Nuclear Regulatory Commission, *PRA Procedures Guide—A Guide to the Performance of Probabilistic Risk Assessments for Nuclear Power Plants*, NUREG/CR-2300, January 1983.
6. Joseph V. Rodricks, Susan Brett, and Grover Wrenn, *Significant Risk Decisions in Federal Regulatory Agencies*, ENVIRON Corporation, Washington, DC, 1987.
7. George Eads and Peter Reuter, *Designating Safer Products: Corporate Responses to Product Liability Law and Regulation*, The Rand Corporation Institute for Civil Justice, Santa Monica, CA, 1983.
8. *Risk Analysis in the Process Industries*, European Federation of Chemical Engineering Publication No. 45, The Institution of Chemical Engineers, Rugby, England, 1985.
9. *Risk Analysis of Six Potentially Hazardous Industrial Objects in the Rijnmond Area; A Pilot Study*, D. Reidel Publishing Company, Dordrecht, Holland, 1982.
10. *Canvey: An Investigation,* Health and Safety Executive, HMSO, 1978. *Canvey: A Second Report*, Health and Safety Executive, HMSO, 1981.
11. C. G. Ramsey, R. Sylvester-Evans, and M. A. English, "Siting and Layout of Major Hazardous Installations," *I. Chem. E. Symposium Series No. 71,* 1982.
12. Vernon H. Guthrie and David A. Walker, *Enterprise Risk Management*, Proceedings of the 17th International System Safety Conference, Orlando, FL, 1999.

13. M. L. Casada, J. Q. Kirkman, and H. M. Paula, "Facility Risk Review as an Approach to Prioritizing Loss Prevention Efforts," *Plant/Operations Progress*, October 1990.

14. *Guidelines for Process Equipment Reliability Data with Data Tables*, American Institute of Chemical Engineers, New York, NY, 1989.

15. *IEEE Guide to the Collection and Presentation of Electrical, Electronic, Sensing Component, and Mechanical Equipment Reliability Data for Nuclear Power Generating Stations*, IEEE Std. 500–1984, The Institute of Electrical and Electronics Engineers, Inc., 1983.

16. OREDA Participants, *OREDA Offshore Reliability Data Handbook (Second Edition)*, DNV, P.O. Box 300, N-1322, Hovik, Norway, 1997.

17. W. Denson et al., *Nonelectronic Parts Reliability Data 1991*, NPRD-91, Reliability Analysis Center, P.O. Box 4700, Rome, NY, 1991.

18. *Failure Mode/Mechanism Distributions 1991*, Reliability Analysis Center, P.O. Box 4700, Rome, NY, 1991.

19. *Systems Reliability Service Data Bank*, National Centre of Systems Reliability, System Reliability Service, UKAEA, Culcheth, England.

20. *Nuclear Plant Reliability Data System: Annual Reports of Cumulative System and Component Reliability for Period from July 1, 1974, through December 31, 1982*, NPRD A02/A03 (INPO 83-034), Institute of Nuclear Power Operations, Atlanta, GA, October 1983.

21. Clayton P. Gillette and Thomas D. Hopkins, *Federal Agency Valuations of Human Life—A Report to the Administrative Conference of the United States (Draft)*, Boston, MA, April 1988.

22. *Guidelines for Hazard Evaluation Procedures, Second Edition with Worked Examples*, American Institute of Chemical Engineers, New York, NY, 1992.

23. *Manual of Industrial Hazard Assessment Techniques*, The World Bank, October 1985.

24. *Guidelines for Chemical Process Quantitative Risk Analysis, Second Edition*, American Institute of Chemical Engineers, New York, NY, 2000.

25. R. E. Knowlton, *Hazard and Operability Studies, The Guide Word Approach*, Chemetics International Company, Vancouver, BC, 1981.

26. John L. Woodward (ed.), *Text Proceedings for the International Symposium on Preventing Major Chemical Accidents*, American Institute of Chemical Engineers, New York, NY, 1987.

27. Steven R. Hanna and Peter J. Privos, *Guidelines for Use of Vapor Cloud Dispersion Models, Second Edition*, American Institute of Chemical Engineers, New York, NY, 1996.

28. John L. Woodward (ed.), *International Conference on Vapor Cloud Modeling*, American Institute of Chemical Engineers, New York, NY, 1987.
29. D. F. Haasl et al., *Fault Tree Handbook*, USNRC, NUREG-0492, Washington, DC, January 1981.
30. F. R. Farmer, "Reactor Safety and Siting: A Proposed Risk Criterion," *Nuclear Safety*, Vol. 8, No. 6, Oak Ridge, TN, November–December 1967, pp. 539–548.
31. "CEFIC Views on the Quantitative Assessment of Risks from Installations in the Chemical Industry," European Council of Chemical Manufacturers' Federations, Brussels, Belgium, April 1986.
32. J. B. Fussell and J. S. Arendt, "System Reliability Engineering Methodology: A Discussion of the State-of-the-Art," *Nuclear Safety*, Vol. 20, No. 5, Oak Ridge, TN, September–October 1979.
33. A. Amendola, "Uncertainties in Systems Reliability Modeling: Insight Gained Through European Benchmark Exercises," *Nuclear Engineering and Design*, Vol. 93, Elsevier Science Publishers, Amsterdam, Holland, 1986, pp. 215–225.
34. F. P. Lees, *Loss Prevention in the Process Industries*, 2nd Edition, Butterworth-Heinemann, Oxford, UK, 1996.
35. D. R. T. Lowe, "Major Incident Criteria—What Risk Should Society Accept?," *I. Chem. E. Proceedings of Eurochem Symposium*, 1980.
36. Bernard L. Cohen and I-Sing Lee, "A Category of Risks," *Health Physics*, Vol. 36, Pergamon Press Ltd., London, England, June 1979, pp. 707–722.
37. Edmund A. C. Crouch and Richard Wilson, *Risk/Benefit Analysis*, Ballinger Publishing Company, Cambridge, MA, 1982.
38. Department of Labor, Bureau of Labor Statistics, Washington, DC, 1997.
39. G. L. Hamm and R. G. Schwartz, *Issues and Strategies in Risk Decision Making*, International Process Safety Management Conference and Workshop, September 22–24, 1993, San Francisco, CA, 351–371, American Institute of Chemical Engineers, New York, NY, 1993.
40. D. C. Hendershot, "Risk Guidelines as a Risk Management Tool," *Process Safety Progress* 15, 4 (Winter 1996), 213–218.
41. J. Pikaar, *Risk Assessment and Consequence Models*, 8th International Symposium on Loss Prevention and Safety Promotion in the Process Industries, June 6–9, 1995, Antwerp, 221–249, Technological Institute of the Royal Flemish Society of Engineers, Antwerp, Belgium, 1995.
42. *Using Layer of Protection Analysis for Estimating Chemical Process Risk* (Final Draft), American Institute of Chemical Engineers, New York, NY, 2000.

43. B. J. M. Ale, "Risk Analysis and Risk Policy in the Netherlands and the EEC," *Journal of Loss Prevention in the Process Industries*, 4 (January), 58–64, 1991.

44. B. J. M. Ale, *The Implementation of an External Safety Policy in the Netherlands*, International Conference on Hazard Identification and Risk Analysis, Human Factors and Human Reliability in Process Safety, January 15–17, 1992, Orlando, FL, 173–183, American Institute of Chemical Engineers, New York, NY, 1992.

45. R. K. Goyal, "Practical Examples of CPQRA from the Petrochemical Industries," *Trans. IChemE*, 71, Part B, 117–23, 1993.

46. Health and Safety Executive (HSE), *Risk Criteria for Land-Use Planning in the Vicinity of Major Industrial Hazards*, London: HMSO, 1989.

47. E. N. Helmers and L. C. Schaller, "Calculated Process Risks and Hazards Management," *Plant/Operations Progress*, 14, No. 4 (October), 229–31, 1995.

48. H. J. Pasman, "Quantitative Risk Assessment in Europe," *Process Safety Progress*, 14, No. 4 (October), 229–31, 1995.

49. F. M. Renshaw, "A Major Accident Prevention Program," *Plant/Operations Progress*, 9, No. 3 (July), 194–97, 1990.

50. K. Whittle, *LPG Installation Design and General Risk Assessment Methodology Employed by the Gas Standards Office*, Conference on Risk and Safety Management in the Gas Industry, October 28, 1993, Hong Kong, 1.1–1.23. Hong Kong: Electrical and Mechanical Services Department of the Hong Kong Government and the Hong Kong Institution of Engineers, 1993.

51. M. J. Pikaar and M. A. Seaman, *A Review of Risk Control*, Zoetermeer, Netherlands: Ministerie VROM, 1995.

52. S. B. Gibson, *The Use of Risk Criteria in the Chemical Industry*, ASME Winter Annual Meeting, November 30–December 1, 1988.

53. T. A. Kletz, "Hazard Analysis–A Review of Criteria," *Reliability Engineering*, 3, 325–38, 1982.

54. Center for Chemical Process Safety, *Tools for Making Acute Risk Decisions with Chemical Process Safety Applications*, New York: American Institute of Chemical Engineers, 1995.

55. V. T. Covello, P. M. Sandman, and P. Slovic, *Risk Communication, Risk Statistics, and Risk Comparisons: A Manual for Plant Managers*, Chemical Manufacturers Association, Washington, DC, 1988.

56. P. Slovic, "Perception of Risk," *Science*, 236, (April 17), 1987, pp. 280–285.

57. R. Wilson and E. A. C. Crouch, "Risk Assessment and Comparisons: An Introduction," *Science*, 236, (April 17), 1987, pp. 267–70.

58. Vincent T. Covello et al. (eds.), *The Analysis of Actual Versus Perceived Risks*, Plenum Press, New York, NY, 1983.
59. William D. Rowe, *An Anatomy of Risk*, John Wiley and Sons, Inc., New York, NY, 1977.
60. William W. Lowrance, *Of Acceptable Risk*, William Kaufmann, Inc., Los Altos, CA, 1976.
61. Ray A. Waller and Vincent T. Covello (eds.), *Low-Probability/High-Consequence Risk Analysis*, Plenum Press, New York, NY, 1984.
62. Chris Whipple, "Redistributing Risk," *Regulation: Journal on Government and Society*, American Enterprise Institute, Washington, DC, May/June 1985, pp. 37–44.
63. *Seven Cardinal Rules of Risk Communication*, OPA-87-020, U.S. Environmental Protection Agency, Washington, DC, March 1996.

SUGGESTED ADDITIONAL READING

Center for Chemical Process Safety, *Guidelines for Chemical Transportation Risk Analysis*, New York: American Institute of Chemical Engineers, 1995.

Center for Chemical Process Safety, *Guidelines for Improving Plant Reliability Through Data Collection and Analysis*, New York: American Institute of Chemical Engineers, 1998.

Covello, Vincent T., Peter M. Sandman, and Paul Slovic, *Risk Communication, Risk Statistics, and Risk Comparisons: A Manual for Plant Managers*, Chemical Manufacturers Association, Washington, DC, 1988.

Gutteling, Jan M. and Oene Wiegman, *Exploring Risk Communication*, Kluwer Academic Publishers, Dordrecht, 1996.

Health and Safety Executive Advisory Committee on Dangerous Substances (HSE), *Major Hazard Aspects of the Transport of Dangerous Substances*, London: HMSO, 1991.

Jones, D., *Nomenclature for Hazard and Risk Assessment in the Process Industries*, Second Edition, Rugby, Warwickshire, UK: Institution of Chemical Engineers, 1992.

Lundgren, Regina E., *Risk Communication: A Handbook for Communicating Environmental, Safety, and Health Risks*, Battelle Press, Columbus, 1994.

Pitblado, R., and R. Turney (eds.), *Risk Assessment in the Process Industries*, Second Edition, Rugby, Warwickshire, UK: Institution of Chemical Engineers, 1996.

Rhyne, W. R., *Hazardous Materials Transportation Risk Analysis*, New York: Van Nostrand Reinhold, 1994.

TNO Institute of Environmental Sciences, *Methods for the Determination of Possible Damage*, Amsterdam: Netherlands Organization for Applied Scientific Research, 1992.

TNO Institute of Environmental Sciences, *Methods for Determining Probabilities*, Amsterdam: Netherlands Organization for Applied Scientific Research, 1997.

TNO Institute of Environmental Sciences, *Methods for the Calculation of Physical Effects*, Amsterdam: Netherlands Organization for Applied Scientific Research, 1997.

TNO Institute of Environmental Sciences, *Guidelines for Quantitative Risk Assessment*, Amsterdam: Netherlands Organization for Applied Scientific Research, 1999.

Wells, G., *Hazard Identification and Risk Assessment*, Rugby, Warwickshire, UK: Institution of Chemical Engineers, 1996.

Publications Available from the CENTER FOR CHEMICAL PROCESS SAFETY of the AMERICAN INSTITUTE OF CHEMICAL ENGINEERS 3 Park Avenue, New York, NY 10016-5991

CCPS Guidelines Series

Guidelines for Process Safety In Batch Reaction Systems
Guidelines for Consequence Analysis of Chemical Releases
Guidelines for Pressure Relief and Effluent Handling Systems
Guidelines for Design Solutions for Process Equipment Failures
Guidelines for Safe Warehousing of Chemicals
Guidelines for Postrelease Mitigation in the Chemical Process Industry
Guidelines for Integrating Process Safety Management, Environment, Safety, Health, and Quality
Guidelines for Use of Vapor Cloud Dispersion Models, Second Edition
Guidelines for Evaluating Process Plant Buildings for External Explosions and Fires
Guidelines for Writing Effective Operations and Maintenance Procedures
Guidelines for Chemical Transportation Risk Analysis
Guidelines for Safe Storage and Handling of Reactive Materials
Guidelines for Technical Planning for On-Site Emergencies
Guidelines for Process Safety Documentation
Guidelines for Safe Process Operations and Maintenance
Guidelines for Process Safety Fundamentals in General Plant Operations
Guidelines for Chemical Reactivity Evaluation and Application to Process Design
Tools for Making Acute Risk Decisions with Chemical Process Safety Applications
Guidelines for Preventing Human Error in Process Safety
Guidelines for Evaluating the Characteristics of Vapor Cloud Explosions, Flash Fires, and BLEVEs
Guidelines for Implementing Process Safety Management Systems
Guidelines for Safe Automation of Chemical Processes
Guidelines for Engineering Design for Process Safety
Guidelines for Auditing Process Safety Management Systems
Guidelines for Investigating Chemical Process Incidents
Guidelines for Hazard Evaluation Procedures, Second Edition with Worked Examples
Plant Guidelines for Technical Management of Chemical Process Safety, Revised Edition
Guidelines for Technical Management of Chemical Process Safety

Guidelines for Chemical Process Quantitative Risk Analysis
Guidelines for Process Equipment Reliability Data with Data Tables
Guidelines for Safe Storage and Handling of High Toxic Hazard Materials
Guidelines for Vapor Release Mitigation

CCPS Concepts Series

Evaluating Process Safety in the Chemical Industry: A User's Guide to Quantitative Risk Analysis
Avoiding Static Ignition Hazards in Chemical Operations
Estimating the Flammable Mass of a Vapor Cloud
RELEASE: A Model with Data to Predict Aerosol Rainout in Accidental Releases
Practical Compliance with the EPA Risk Management Program
Local Emergency Planning Committee Guidebook: Understanding the EPA Risk Management Program Rule
Inherently Safer Chemical Processes. A Life-Cycle Approach
Contractor and Client Relations to Assure Process Safety
Understanding Atmospheric Dispersion of Accidental Releases
Expert Systems in Process Safety
Concentration Fluctuations and Averaging Time in Vapor Clouds

Proceedings and Other Publications

Proceedings of the International Conference and Workshop on Modeling the Consequences of Accidental Releases of Hazardous Materials
Proceedings of the International Conference and Workshop on Reliability and Risk Management, 1998
Proceedings of the International Conference and Workshop on Risk Analysis in Process Safety, 1997
Proceedings of the International Conference and Workshop on Process Safety Management and Inherently Safer Processes, 1996
Proceedings of the International Conference and Workshop on Modeling and Mitigating the Consequences of Accidental Releases of Hazardous Materials, 1995
Proceedings of the International Symposium and Workshop on Safe Chemical Process Automation, 1994
Proceedings of the International Process Safety Management Conference and Workshop, 1993
Proceedings of the International Conference on Hazard Identification and Risk Analysis, Human Factors, and Human Reliability in Process Safety, 1992
Proceedings of the International Conference and Workshop on Modeling and Mitigating the Consequences of Accidental Releases of Hazardous Materials, 1991
Safety, Health and Loss Prevention in Chemical Processes: Problems for Undergraduate Engineering Curricula